# PHANTOM SOLDIER

# PHANTOM SOLDIER

## THE ENEMY'S ANSWER TO U.S. FIREPOWER

*ILLUSTRATED*

H. JOHN POOLE

FOREWORD BY
WILLIAM S. LIND

POSTERITY
PRESS

Published by Posterity Press
P.O. Box 5360, Emerald Isle, NC 28594
(www.posteritypress.org)

*Cataloging-in-Publication Data*
Poole, H. John, 1943-
Phantom Soldier.
    Includes bibliography and index.
    1. Infantry drill and tactics.  2. Firepower.
    3. Military history.  4. Military art and science.
    5. Maneuver warfare.  6. United States Marine Corps.
I. Title.      ISBN 10: 0-9638695-5-8      2001      355'.42
              ISBN 13: 978-0-9638695-5-5
Library of Congress Control Number          2001-130548

Cover art © 2001 by Michael Leahy
Edited by Dr. Mary Beth Poole
Proofread by Martha B. Spencer


Fourth printing, United States of America, July 2006

*This book is dedicated to the Americans who must fight the next war. May the lessons of the past help them to better prepare for the future.*

# Contents

# Illustrations

**Maps**

**Figures**

# Foreword

Throughout most of the modern era, any non-Western military force that wanted to fight Western armed forces had to copy them. It had to adopt Western military discipline, tactics, training, and technology. Failure to do so meant inevitable defeat, as the Chinese, among others, found over and over again in the 19th century.

But this is no longer true. In the second half of the 20th century, a new pattern began to emerge: when Western armed forces fought non-Western opponents, they lost. The French were defeated in Vietnam and in Algeria, the Soviets in Afghanistan, and the Americans in Vietnam, Lebanon, and Somalia. The Israeli Defense Forces, once the best "Third Generation," maneuverist armed forces in the world, were defeated in Lebanon by Hezbollah and are currently being defeated again by the Intifada. As we move into a world not just of nations but of cultures in conflict, the implications of this vast sea change are profound.

In this book, John Poole explains one non-Western way of war, the Oriental way of war. He does so not in academic or theoretical terms, but practically, in terms of small-unit tactics and techniques — the sort of thing NCOs and junior officers need to know. A former Marine Staff NCO himself, John Poole understands what is important in enabling others like him to stay alive and prevail in small-unit combat.

Soldiers and Marines in Western countries need to understand the Oriental (and other non-Western) ways of war for two reasons. First, they are likely to come up against it. For no good strategic reason, the United States and China seem to be on a collision course. The Chinese now have and know how to employ modern weapons. But their techniques, tactics, and strategies are not likely to be simple copies of those used by the West. They have a long military tradition of their own, and it leads them in some fundamentally different directions. The better we understand the differences beforehand, the fewer lessons we will have to learn in combat, the hard way.

Second, as we study the Oriental way of war, in this book or elsewhere, we may find that, at least in some cases, their approach makes more sense than our own. This is especially true for a Second Generation military, which is what the U.S. Armed Forces largely remain, official U.S. Marine Corps doctrine to the contrary (as Marines often put it, what the Marine Corps says is great, but it's not what it does). Second Generation warfare reduces war to little more than the methodical application of firepower to destroy targets. The Oriental way of war is far more sophisticated. It plays across the full spectrum of conflict — the moral and mental levels as well as the physical. Even at the physical level, it relies on the indirect approach, on stratagem and deception, far more than on simple bombardment. Seldom do Asians fall into mindless *Materialschlact* or "body counts"; and while Oriental armies often can (and have) taken many casualties, their tactics at the small-unit infantry level are often cleverly designed to spare their own men's lives in the face of massive Western firepower.

A truly professional military is always looking for better ways to do things. One of the most common sources of new and better ideas is foreign practice. Between the World Wars, American military journals were full of articles on foreign tactics and techniques. Sadly, that is no longer the case. Perhaps because of the delusion that wars are won by the side that has the most complex technology, we have largely stopped trying to learn from others.

This book offers an important, perhaps a life-saving, opportunity to reverse our short-sighted current practice. It presents the Oriental way of war in depth, but also understandably. If official field manuals remain largely unimaginative and uninspired, there is no reason squad leaders and platoon and company commanders must let their own tactics and techniques be set-piece and predictable. Here, as in his previous books, John Poole offers a better way.

> — William S. Lind
> author of *Maneuver Warfare Handbook*

# Preface

When Marco Polo returned from the Far East to Venice in 1295, he warned of the Karauna raiders in the barren deserts of Upper Persia. Of Indian and Mongolian descent, the Karaunas were said to have the power to conjure up a magic, choking gloom through which to approach lucrative caravans.[1]

> In India they acquired the knowledge of magical and diabolical arts, by means of which they are enabled to produce darkness, obscuring the light of day to such a degree, that persons are invisible to each other, unless within a very small distance. Whenever they go on their approach is consequently not perceived.[2]
> — Marco Polo

Marco Polo's perception may have been clouded by superstition and sandstorm, but those with whom he had lived in China firmly believed in mystical occurrences.[3]

> Wind can be acquired by sacrifice, rain can be obtained by prayer, and cloud, fog, thunder and lightening can be summoned by conjuration.[4]
> — Tou Bi Fu Tan
> *A Scholar's Dilettante Remarks on War*

The public-television documentary — *All the Kings' Men* — describes a hauntingly similar occurrence at the beginning of the 20th century. It would seem that the British Sandringham Company vanished forever after advancing into a cloud of "golden mist" at Gallipoli in 1915. The documentary concludes that the Turks must have somehow used the mist to conceal a massacre.[5] While the Allied soldiers' fate may be hypothetical, the mist wasn't.

According to many writers . . . , hundreds of British troops were mysteriously "abducted" by a cloud that settled over them as they advanced toward the Turkish positions during one of the battles of the Gallipoli campaign in 1915. The source for this story is a statement written 50 years after the incident by three New Zealand soldiers, who deposed that they had watched a dense, solid-looking cloud shaped like a loaf of bread settle on the ground in the path of an advancing column of troops. After the men walked into it, the account went on, the cloud lifted, leaving no one behind.

In *Into Thin Air,* Paul Begg concluded that this disappearance could not have happened as described. The battalion named by the New Zealanders was not unaccounted for. Another battalion had been decimated in battle, but that was nine days before the date given in the statement, and the report of the postwar commission that investigated the disastrous Gallipoli campaign included mention of an unseasonable mist that blinded Allied artillery gunners but aided their Turkish counterparts in wiping out a British unit. Significantly, that report was only fully declassified in 1965 and its publication may have brought forth a confused recollection by the New Zealanders.

Although the details are questionable, a mystery concerning the fallen still remains. As Begg notes: "Of the 34,000 British and Empire troops who died at Gallipoli, 27,000 have no known grave."[6]

More recently, Karen rebels from Myanmar (formerly Burma) have claimed adolescent leaders who can read minds, smell distant foes, and become invisible.[7] Though farfetched, their belief is nonetheless useful to understanding Oriental military tradition. Still available in Buddhist bookstores is an ancient document providing its reader with a step-by-step procedure on how to appear not to exist.[8]

Whether the Karaunas, Turks, or Htoo brothers could really disappear from view is not at issue. All that must be acknowledged is that Eastern soldiers have long tried to disguise their presence on the battlefield. As this book will show, many have managed to do just that during some of the fiercest fighting in U.S. history. On offense, they have sprung from nowhere to damage targets of strategic importance. On defense, they have tactically withdrawn while pretending to succumb to bombardment or languish below ground. Though critically short of armor and air support, they have suffered far fewer casualties than previously thought. To survive the more lethal weaponry of the 21st century, U.S. infantrymen must do what the "phantom soldiers" did.

# Acknowledgments

To the United States of America and its Armed Forces goes a heartfelt thanks for the opportunity to serve. To the Good Lord goes the credit for any insight into how to minimize the loss of life in war.

From May 1997 through February 2001, 30 Marine battalions (20 of them infantry) and one Naval special-warfare group have participated in multiday training sessions on the squad techniques in *The Last Hundred Yards: The NCO's Contribution to Warfare* and the enemy overview in *One More Bridge to Cross: Lowering the Cost of War*. Their receptivity to the more productive — "bottom-up" — way of training and operating has inspired the writing of this book. *Semper fidelis.*

# Part One

## The Eastern Way of War

When you are ignorant of the enemy but know yourself, your chances of winning and losing are equal. — Sun Tzu

# Unheralded
# Success

● *How well have America's Eastern adversaries fared?*

● *How good were their small-unit infantry tactics?*

(Source: *Handbook on the Chinese Communist Army*, DA Pamphlet 30-51 (1960), U.S. Army, figure 110; Corel Gallery, *Clipart* — Plants, 35A005)

## A Land Shrouded in Mystery

To many Americans, Asia is still a place of simple peasantry and agrarian lifestyle. Having fallen behind the Western World technologically, its peoples are thought to be behind in every other aspect of human endeavor. But preconceived notions can be dangerous. This gap between East and West in science and engineering has not always existed. Over 4500 years ago, while Northern Europeans groveled in caves, the inhabitants of the Indus Valley enjoyed public baths and flowing sewers in the sister metropolises of Mohenjo-daro and Harappa.[1] At the start of the last millennium, the Chinese army was the world's most advanced technologically. It had iron-tipped arrows, crossbows, gunpowder, land mines, flamethrowers, and chemical weapons. It was soon to have multistage rockets and cannons.[2]

It is difficult to ascertain how long the Asians managed to maintain this edge in weaponry. Inside the Jaigarh Fort just north of Jaipur, India, sits a cannon that in 1720 could shoot an explosive shell 22 miles[3] — close to maximum range for modern howitzers. At beginning of World War II (WWII), Japanese ships, planes, and munitions were roughly equivalent to those of the United States. Then, the Eastern nations began to fall further and further behind in weapons development. How, in the ensuing decades, did their armies still manage to project so much military power?

The answer is quite simple. Only through maneuver could they maximize the effect of their own "low-tech" weapons and minimize the effect of the opposition's "high-tech" weapons. So they compensated by enhancing their small-unit infantry skills. Behind the Bamboo Curtain, their methodology came to be known as "low-tech people's war." Just 30 years ago in the lowlands of Southeast Asia, the world got to see once again what this alternative style of fighting can accomplish. It was there that an invasion force, with dangerously extended supply lines, needed no tanks or planes to fight the world's most technologically advanced nation to a standstill. To appreciate the power of small-unit technique, one must only review how the Germans almost salvaged World War I (WWI).

### Many an Eastern Foe Has Fared Better Than Reported

Every German infantry squad that participated in the Spring offensives of 1918 had been trained in "stormtrooper" squad assault technique.[4] This technique permitted a single squad to safely attack successive rows of Doughboys behind barbed wire and machineguns. The German soldiers had only to follow a simple sequence: (1) crawl up to the protective wire, (2) emplace a bangalore torpedo, (3) hit the target trenches with precision artillery fire, (4) blow the bangalore, and (5) transit the breach when the fire stopped. By using bayonets and grenades, the stormtroopers could keep the defenders of adjacent sectors from realizing that a ground attack was underway. By that stage of the war, there had been so many echelons of U.S. infantrymen sent forward that the German squads could not penetrate deep enough to get behind them all.

In the spring of 1918, the Germans embarked on a major offensive. . . . The new offensive doctrine called for . . . assault units to infiltrate enemy weak spots and drive deep into his rear to disorganize his defense. The tactics worked well. . . . Modern infantry tactics are rooted in German defense-in-depth and infiltration tactics.[5]
— MCI 7401, U.S. Marine Corps

On the other side of the world, Japan had become interested in protecting its most sophisticated weapon — the individual infantryman — after suffering tremendous losses in its 1904-05 victory over the Russians.

By 1920 IJA [Imperial Japanese Army] tacticians realized the need to disperse infantry formations in order to reduce losses when attacking a defender who possessed the lethal firepower of modern weapons. The revised 1925 edition of the *Infantry Manual* emphasized tactics designed to allow the attacker to reach the enemy defender's position.[6]
— *Kwantung Army*

The IJA founded its battle doctrine on bold offensive operations. . . . The IJA relied on the infantry as its main battle force. . . . [T]acticians had to guarantee that the attacking Japanese infantry reached the enemy positions with a minimum of friendly losses.[7]
— Leavenworth Papers No. 2, U.S. Army

By 1939, the Japanese had come close to perfecting their individual and small-unit tactical technique. Still they failed to secure victory in their under-reported, six-month battle with the Russians at Nomanhan, Manchuria. Their nemesis had field-wise infantrymen, better logistics, and more armor. Though second-to-none tactically, the Japanese had been operationally outmaneuvered by Soviet armor.

Japanese strength . . . lay in small units and the epitome of Japanese doctrine was embodied in small-unit tactics. Night attacks . . . and the willingness to engage in hand-to-hand combat were the hallmarks of the Japanese infantryman. Indeed such tactics were very successful against Soviet infantry [in 1939].[8]

> — Leavenworth Papers No. 2, U.S. Army

The Japanese were soon forced to face the same reality the Germans had in WWI. No amount of infantry expertise can stop an opponent willing to invest any number of men and machines. The Allied forces that pushed the Japanese back from Imphal, India, during WWII had not only vastly more supplies/equipment, but also eight times more men.[9]

> Coupled with the spiritual or psychological values of offensive spirit . . . such tactics produced one of the finest infantry armies in the world. . . . It [the IJA] was an army . . . that tried to use doctrine to compensate for materiel deficiencies.[10]
>
> — Leavenworth Papers No. 2, U.S. Army

Then, in Korea, this strategy of massing men and machines didn't work out as well for Western forces. In a few short months in late 1950, poorly supplied and largely equipmentless Chinese were able to push U.N. troops back from the Yalu River in the east and Chosin Reservoir in the west to south of 38th Parallel. Somehow the power differential between Eastern- and Western-style armies had shifted.

> I'm afraid we haven't recognized the most important lesson from Korea. The Communists have developed a totally new kind of warfare. . . . This is a total warfare, yet small in scope, and it's designed to neutralize our big . . . weapons. Look at Vietnam. The French outnumbered the Communists two to one, yet they [the French] were massacred.[11]
>
> — Lt.Gen. Chesty Puller USMC (Ret.)

## There May Be Something Here to Learn

Perhaps, instead of discrediting the "technologically inferior" Eastern armies, Western armies should try to learn from them how to overcome a temporary deficit in firepower. Within "low-tech" people's army tactical techniques may lie the answer to fewer U.S. casualties.

There may be some Americans who for political or economic reasons prefer the status quo. They will quickly point out that Eastern armies have suffered "appalling" losses in their wars with the United States. But one must remember that it is primarily by U.S. approximation that those losses have been so heavy. Body counts can be misleading. Depending on the urgency and method with which they are made, they can finally encompass every alien corpse (whether in uniform or not), blood trail, diving opponent, and unaccounted-for defender. To arrive at an accurate total, one must minimally do two things: (1) refrain from including every known member of units from which a few bodies are found, and (2) make sure that each body has only been reported by one observer.

Occasionally in a U.S. publication, a retired Asian military officer will admit to huge losses in some long-ago war. Each admission must be carefully analyzed. If the author was a frontline infantry commander, it may focus on the least successful of many campaigns. If he was a support unit commander or high-level staff officer, it may incorporate noninfantry casualties from interdiction bombing. If he was a guerrilla unit advisor, it may address civilian sympathizers killed by U.S. supporting arms. If he was a political officer or cadre member, it may be a subtle attempt to draw attention away from his country's wartime advantage. In a totalitarian nation, statistically exacerbating the arrogance of a traditional foe would evoke few repercussions and possibly a medal.

Moreover, even comparing the casualties actually suffered by opposing infantry units will not necessarily reveal their relative proficiency at individual and small-unit combat — the key to soldier survivability. The casualty totals have been influenced by how much help each unit got from its respective support establishment. While incurring whatever part of the "appalling" overall losses in that long-ago war, the Eastern infantry units were probably operating with few supplies, no air cover, and no tank reinforcement. With more backing, their part of those losses would have been far less. If — while so disadvantaged technologically — they could still hold their own tactically, then — with equivalent logistics and fire support — they might win with very few casualties whatsoever.

## What Might Be That Eastern Advantage?

Infantry knowledge does not appear to build upon itself like other types of knowledge. As weapons become more lethal, combat must necessarily be conducted by smaller, more isolated groups of soldiers. What the lower ranks have learned about combat has had great difficulty entering the institutional knowledge of the British, French, and American infantry. Therein may lie one of the major differences between West and East. An Oriental army places the highest priority on the training, informing, and utilization of the individual soldier and small unit. It assumes that, with enough skill and leeway, human beings can outsmart machines.

> Whether in guerrilla . . . or limited regular warfare, waged artfully, it [armed struggle] is fully capable of . . . getting the better of a modern army like the U.S. Army. . . . This is a development of the . . . military art, the main content of which is to rely chiefly on [the] man, on his patriotism and . . . spirit, to bring into full play all weapons and techniques available to defeat an enemy with up-to-date weapons and equipment.[12]
> — General Vo Nguyen Giap
> *The Military Art of People's War*

## The Lack of Resources Has Spawned Infantry Excellence

A nation does not need a dynamic political system to have a strong military. While fascist and communist countries have totalitarian governments, not all have authoritarian armies. In fact, some may permit more democratic processes than those in the Free World. Several dictators have confirmed this paradox.

> Apart from the role played by the Party, the reason why the Red Army can sustain itself [during the period of the second revolutionary war] without collapse in spite of such a poor standard of material life and such incessant engagements, is its practice of *democracy* [italics added]. The officers do not beat the men; officers and men receive equal treatment; soldiers enjoy freedom of assembly and speech; cumbersome formalities and ceremonies are done away

with; and the account books are open to the inspection of all.[13]
— Mao Tse-tung in 1928

It would appear that all but the last of America's 20th-century adversaries have applied (in varying degrees) this more "democratic" approach to military operations. This has permitted their better adapting to the decentralized nature of modern combat. By the end of 1917, the Germans had put their noncommissioned officers (NCOs) in charge of not only the autonomous forts that comprised their matrices of defense,[14] but also the maneuver elements that spearheaded their ground attacks.[15] In World War II, the German *Volksgrenadier* (people's infantry) divisions proved particularly effective in the defense of the Rhineland and Berlin.[16]

It had taken just sixteen days for the troops of three fronts — the First Byelorussian, the First Ukrainian and the Second Byelorussian — to crush enemy forces in the area of Berlin and to seize the German capital. . . . [T]he cost of the final battle in Soviet casualties — killed, wounded, missing — was enormous: 305,000 from April 16 [1945] to May 8.[17]
— Georgi K. Zhukov

Of course, the Germans were not the only ones to discover the benefits of decentralized control over what amounts to a highly compartmentalized and fast-moving situation. So too had their pre-WWI allies since 1872 — the Japanese.[18]

These [improvements to the Imperial Japanese Army of 1939] . . . included reliance on the independent decision-making ability of junior officers and non-commissioned officers.[19]
— Leavenworth Papers No. 2, U.S. Army

As Japan had acquired much of its culture from Mongol invaders, T'ang Dynasty emissaries, and Buddhist missionaries,[20] its mainland cousins may have long ago realized the same paradox. Perhaps China, Korea, and Vietnam had all succeeded in delegating tactical-decision-making authority to those in direct contact with the enemy.

> Those who understand big and small units will be victorious.[21]
>
> — Sun Tzu

Sun Tzu advocated that one should take the initiative and be flexible in war.[22]

When the rapid-fire machinegun removed much of the appeal from massed frontal assault, the Chinese must have put Sun Tzu's theories into wider practice. U.S. Marine raider battalions patterned after Mao Tse-tung's 8th Route Army helped to turn the tide at Guadalcanal.

> Colonel Carlson had some very specific ideas about training his men, and one was to indoctrinate them with the same sort of democratic spirit as that practiced in the 8th Route Army. Their was very little distinction between officers and men, and the slogan for the 2nd Raider Battalion was "gung ho," or "work together," a slogan borrowed directly from the Chinese communists.[23]

> Evans Carlson introduced all the troops to the 8th Route Army system: democracy in the ranks. First he invited them to criticize him and the other officers at will. That was how Mao Zedong [Tse-tung] and Jue De did it, and it had worked, for Carlson had seen it work.[24]

By 1950, the major armies of mainland Asia may have given their frontline troops more of a say on which tactical techniques to develop and rehearse. In determining what to do next in active combat, Eastern commanders may have gone so far as to occasionally rely on the consensus opinions of their squad and fire-team leaders (NCOs). Discipline did not appear to suffer in the process.

> The Chinese troops were a unique enemy, without any of the characteristics of a traditional American or European army. There were no officer corps or ranks in this "people's army." The soldiers addressed their commanders as "Comrade Platoon Leader" or "Comrade Company Commander" and were informed in great detail of their tactical and strategic missions.[25]

Then in 1955, the Chinese Army adopted a system of personal rank,[26] but the associated rights and responsibilities were quite different from those in the Western World. Political officers helped to ensure that loyalty to the nation took precedence over that to a unit. Commanders were forced to become familiar with their "subordinates."

> Commanders and political officers above company level and staff officers, specialists, and administrative cadres are required to spend several months each year as privates or as platoon or command officers. . . . This program is regarded as both political and military training, intended to improve officer-enlisted relations and to familiarize those going "back to the ranks" with current problems, activities, and attitudes at basic levels in the armed forces.[27]
> — *Handbook of the Chinese Communist Army*
> DA Pamphlet 30-51, 7 December 1960

Though badly outgunned, the Chinese had gained a stalemate in Korea and the Viet Minh, a victory in Vietnam. Then in 1975, the power of the less bureaucratic, "bottom-up" approach to warfare became impossible to ignore. After "having never lost a battle," the powerful U.S. military came home. To the democratic processes within the North Vietnamese Army [NVA] goes much of the credit for the concerted effort. Most of the NVA units had been forced by attrition to do without political officers.[28]

> In *Kiem thao* sessions, the [NVA] soldiers offered judgments of their comrades [and commanders] and listened to evaluations of their own performances. The meetings sometimes featured discussions of tactics from the unit's recent engagements or *suggestions . . . from the army command* [italics added].[29]

## Combining Infantry Excellence with Better Fire Support

"Low-tech" people's war and "high-tech" mobile-war capabilities are not mutually exclusive. In fact, they complement one another. The world got to assess what hastily trained and underequipped infantry can do with state-of-the-art fire support in December of 1944. That's when *Volksgrenadier* divisions,[30] backed by the latest German tanks and planes,[31] entered the Bulge.

With enough delegated authority to call and adjust aircraft and artillery strikes, small infantry units could operate deep in enemy territory with relative impunity.


## China Is Once Again on the Move

Now at the dawn of the 21st century, China is trying to reclaim its "former position at the center of the civilized world." Unfortunately, with the Korean incursion of 1950, brutal annexation of Tibet in 1951, war with India in 1962, punitive expedition into North Vietnam in 1979,[32] and border clashes with the Soviet Union, past regimes have demonstrated expansionist leanings. Just since 1996, a Maoist uprising in Nepal has claimed over 2000 lives.[33] While traveling across the Indian subcontinent in 1999, a retired U.S. Marine noticed how happy, well fed, and gainfully employed the Nepalese people were in comparison to their southern neighbors. He could find no internal justification for the uprising.[34] There is evidence that China expects its sphere of influence to encompass Japan and all of Southeast Asia within the next ten years.[35]

To become the next superpower, China needs only to better its aerial sensors and supporting arms. To fully appreciate its potential for technological growth, one must only view, firsthand, the modernistic skyscrapers and bustling commerce of present-day Shanghai.[36] Sadly, some of this growth may be channeled toward other-than-peaceful purposes.

China has continued to use its military to reassert what it sees as its sovereignty over Taiwan and the Spratly Islands. Twenty years of economic reform have created the competitive spirit and cash flow on which military establishments thrive. Already there have been improvements to China's sea, air, and missile capabilities. First tested in August 1999, the long-range Dongfeng-31 ballistic missile could be Beijing's answer to any missile defense system that the U.S. might deploy in the region. More agile than its predecessors, the multiple-warhead Dongfeng can better penetrate an antimissile curtain. With these and other technological advances, the Chinese believe that they can regain their lost glory as the greatest military power in Asia, if not the world.[37]

## What Appears to Differentiate the Styles of War

When it comes to a subject as complicated as warfare, generalizations are risky. Each nation develops its own way of fighting based on unique circumstances. Still there are trends. If one were to summarize the differences between Eastern- and Western-style armies, one might say that the former generally do a better job of harnessing the perceptions and common sense of the people in contact with the enemy. Deceptive and multifaceted, this alternative "style of war" is difficult for the Western, "top-down" thinker to comprehend. At times, it employs massive firepower; but more often, it relies on surprise. Its essence lies not in established procedure, but rather in flexibility to change. It encourages its practitioners to shift rapidly between opposites — to alternately use one maneuver as a deception and its reflection as a follow-through.

(Source: Corel Print House, *Plants*, Tree 9)

During the first half of the 20th century, the tactical successes of armies using the Eastern method have been overshadowed by their overall defeat. Only with the Korean stalemate of 1953, did the U.S. public first start to suspect that the Eastern way of war had substance. Now, 25 years after the unsettling end to the Vietnam conflict, few Americans continue to blame Congress. Many who were there believe the problem to have been one of overcentralized control and outmoded small-unit infantry tactics.

> [S]ince the tragic, inevitable fall of Saigon, there has been no major, honest post mortem of the war. There have been critiques dealing with the big picture . . . but none has addressed the lessons learned the hard way, at the fighting level, where people died and the war was in fact, lost.[38]
> — Col. David H. Hackworth U.S. Army (Ret.)

If U.S. citizens were now to more closely examine the Eastern style of war and possibly even demand its assimilation, they might still give meaning to the 58,000 American lives lost in Vietnam. There is evidence that this alternative way of fighting is not only less risky, but also more moral. Without the political subversion and summary executions that have so often accompanied it, the Asian method shows great promise. But these are emotionally charged issues that need not be resolved here. What is important is that the current crop of American servicemen and women come to better understand potential adversaries. Of the world's major armies, only the French, British, and American have yet to develop something similar to the Eastern style of war. Few, if any, Americans who fought at Belleau Wood, Bataan, the Chosin Reservoir, or Khe Sanh were adequately briefed on their opposition. God only knows, how many more could have come home if they had been. So, in the interests of preserving America's greatest asset — its youth — the following treatise on Oriental war is respectfully submitted.

# Strategic
# Advantage

- *How do Eastern and Western thought processes differ?*
- *What gives the Easterner the strategic advantage?*

(Source: Corel Gallery, *Clipart* — Man, 28V011; Plants, 35A021)

## The Origins of Oriental Military Philosophy

To fully appreciate the Asian's paradoxical approach to war, one must be well acquainted with his spiritual roots. Sun Tzu's tactical beliefs appear to have taken much of their inspiration from older religious philosophies. By searching for "ultimate reality," most Eastern religions implicitly promote problem solving through common sense. The most popular — Buddhism — originated in northeastern India around 550 B.C. and then quickly spread into southern China. By 700 A.D., it had been widely accepted throughout Southeast Asia, Tibet, China, Mongolia, Korea, and Japan.[1] Early Buddhism was not a religion in the sense that it promoted deity worship, but rather a way of life. It preached overcoming adversity through the "eightfold path." Right Action alluded to selfless activity. Right Effort was the product of will power. Right Mindfulness (or awareness) required self-examination — constantly reassessing one's behavior and then removing the causes of misdeeds. Right Concentration (or meditation) involved the search for truth — pondering often and deeply on ultimate reality.[2]

Sun Zi (Sun Tzu) made no attempt to remove war from the moral, political and diplomatic arena. He must have realized that victory can best be served by the realistic, timely, and ethical solution to each set of battlefield circumstances.

> Sun Zi grasped the quintessence of war and the relationship between the outcome of a war and the political, economic and diplomatic factors of a society.[3]

He may have also foreseen the problems associated with a long chain of command. Far from the actual fighting, a higher head-quarters might easily misinterpret conditions and require excessive force. Perhaps that's why Sun Zi advised his followers to "understand both big and small units,"[4] and whenever possible "to subdue the enemy without fighting."[5] Contrary to popular opinion, his disciples have long sought the middle ground between military proficiency and moral integrity.

## Minimizing the Effects of War

With the best of intentions, Western wars have often been waged with too little regard for their political and diplomatic ramifications. In deciding to bomb the public utility and transportation apparatus in the Kosovar region, for example, NATO planners may have placed too low a value on the economic infrastructure of a diverse society. It will be quite some time before the tempo of commerce forces its ethnic factions to cooperate.[6] The Eastern world has made no such attempt to separate mutually dependent variables. While its use of political subversion and summary executions can never be condoned, its attempt to project less overall force deserves a closer look. It was through a ghostlike invasion of South Korea in 1950, that the North Korean army left that nation's populace and infrastructure relatively intact. The Allied counteroffensive of early 1951 proved much less successful in this regard.

> The North Koreans were like ghosts. They passed over the countryside and left no mark on it in many ways. But when you use the rock crusher techniques of an American army you hurt your friends. And that was true in Vietnam as well as in Korea.[7]
> — Maj.Gen. Edwin Simmons USMC (Ret.)
> former director of History and Museums, HQMC
> PBS Special *Korea — the Unknown War*

As the Asian has had fewer resources with which to pursue his war aims, he has learned how to better preserve those resources. He knows how to confiscate enemy supplies and live off the land.

> Resourceful generals must try every means to live off the enemy.[8]

The Asian greatly values what his individual soldiers and small units can accomplish. While trying to minimize their overall losses, he seldom overprotects them in each engagement. He may have slightly different rationale for "civilized behavior." During the Chinese Wars of Revolution, the Reds encouraged the humane treatment of enemy soldiers as a public-relations and recruiting ploy.

> In the propaganda directed to the enemy forces, the most effective means are releasing the captured soldiers and giving medical treatment to their wounded. . . . This immediately shatters the enemy's calumny that the Communist bandits kill everyone on sight.[9]
> — Mao Tse-tung

## A Different Thought Process

While Sun Zi's *Art of War* formed the basis for Eastern military thought, Clausewitz's *On War* did the same for the West. Their subtle differences show the Easterner to enjoy a strategic advantage.

The former [the *Art of War* element diagram] indicates the whole governing the parts, while the latter [the *On War* element diagram] implies the parts integrating into the whole.[10]

It is, after all, through pursuing overall strategy that wars are won. In the West, actions against enemy forces (tactics) help to advance operations, and operations to further strategic goals. However, every battlefield action of a Western army does not contribute to its strategic progress. It would appear that Eastern combatants have more carefully invested their tactical effort. This is due, in part, to their long heritage of maneuver (as opposed to attrition) warfare. Also different is the way in which Asians solve problems.

There is, however, much difference between the East and the West in cultural heritages, in values, and in ways of thinking. In the Eastern way of thinking, one starts with the whole, takes everything as a whole and proceeds with a comprehensive and intuitive synthesization [combination]. In the Western way of thinking, however, one starts with the parts, takes [divides] a complex matter into component parts and then deals with them one by one, with an emphasis on logical analysis. Accordingly, Western traditional military thought advocates a direct military approach with a stress on the use of armed forces.[11]

While challenged politically, Mao Tse-tung was nevertheless a brilliant tactician. He understood that the relationship between tactics and strategy could be nonlinear.

The view that strategic victory is achieved by tactical successes alone is erroneous. . . .

. . . In a war, some defeats or failures in battles or campaigns do not lead to a change for the worse in the whole military situation because they are not defeats of decisive significance. . . .

As to the relationship between the whole and the parts, it holds not only between war strategy and operational direction but also between operational direction and tactics. The relation between the action of a division and that of a

regiment or battalion, and the relation between the action of a company and that of a platoon or squad, are concrete illustrations. The commanding officer at any level should centre his attention on the most important and most decisively significant problem in the whole situation he is handling, and not on any other problems or actions.[12]
— Mao Tse-tung

## In the East, Strategy Drives Everything

Every Eastern soldier of whatever rank is kept thoroughly informed of his parent units' overall plans. No subordinate action, however small — whether tactical or otherwise — is contemplated outside this context. That action must do one of two things: (1) further friendly strategy, or (2) attack enemy strategy.[13] By attacking the enemy's strategy, victory can often be won before the battle starts.[14] From this comes the concept of "decisive" engagement — one that furthers the overall war effort.[15] Of course, Western commanders like to think that strategy drives everything their armies do as well. The difference is in how the strategy is applied to the myriad of small-unit engagements. In the West, subordinate units are issued detailed instructions on how to participate in an overall scheme of maneuver. In the East, every member of every unit is carefully briefed on the overall plan and then allowed an opportunity to help. Subordinate-unit commanders get to choose not only how to further the overall war effort, but also how to handle unforeseen circumstances.

The importance or decisive significance of a thing is not determined according to circumstances in general or in the abstract, but according to the concrete circumstances.[16]
— Mao Tse-tung

The West has operated largely in the abstract, relying on the wisdom and leadership of its generals. The East has paid more attention to concrete circumstances, depending on the observations and common sense of the people in direct contact with the enemy. This is readily apparent from how the two sides operate in war. While the West doggedly applies what it has learned about itself in peacetime, the East flexibly applies what it learns about itself and its adversary during the actual fighting.  One becomes progressively more aware of the changing circumstances of war, while the other does not.

> Reading books is learning, but application is also learning and the more important form of learning.  To learn warfare through warfare — this is our chief method.[17]
> — Mao Tse-tung

This combination of strategic direction and grass roots tactics creates a moral amalgam quite different from the normal stereotype. If Asian ground troops cannot make a strategic contribution, they are not committed.  On offense, short-range infiltrators and spies provide enough information about the defender's situation to provide this level of assurance.  On defense, "tactical withdrawal"

(Source:  Courtesy of Cassell PLC, from *World Army Uniforms since 1939,* © 1975, 1980, 1981, 1983 by Blandford Press Ltd.)

and "strategic retreat" provide sufficient opportunity to create more favorable circumstances. The Easterner will not switch to the counteroffensive until a number of conditions have been met.

> We [the Chinese] cannot switch to the counteroffensive unless we have secured during the retreat two of the conditions listed below. . . .
> 1. The people give active support to the Red Army;
> 2. The terrain is favorable to operations;
> 3. The main forces of the Red Army are completely concentrated;
> 4. The weak spots of the enemy are discovered;
> 5. The enemy is worn out both physically and morally; and
> 6. The enemy is induced to make mistakes.[18]
> — Mao Tse-tung

Just to locate a target of strategic import, a Western unit will attack on line and exploit breakthroughs. It has little experience with strategic retreat or "soft" defense. As the Eastern unit does less fighting, it can better preserve resources. It places fewer soldiers and noncombatants (from either side) into harm's way.

> The best way to prevent one's troops from getting exhausted and their ardor from becoming dampened, and to forestall commotion is to "subdue the enemy without fighting."[19]

Again the relative emphases of Sun Zi and Clausewitz reveal how the two ways of war differ. While the Eastern commander avoids combat wherever possible, his Western counterpart seeks it out.

> If it is not in the interests of the state, do not act. If you are not sure of success, do not use troops. If you are not in danger, do not fight a battle.[20]

> First, Sun Zi emphasized the employment of stratagem, on subduing the enemy without fighting, that is, achieving the best possible results at the least possible cost. . . . Clausewitz, however, emphasized putting out of action enemy forces through combats.[21]

Additionally, the Eastern commander puts more emphasis on when, where, and with what to fight — i.e., on the situation. However charismatic, he does not ask subordinates to do more than they can reasonably be expected to accomplish.

> Secondly, Sun Zi [Sun Tzu] emphasized taking advantage of the [combat] situation. He observed, "He who takes advantage of the situation uses his men in fighting as rolling logs or rocks. . . . The side which possesses the momentum of round rocks tumbling quickly from a mountain has victory in hand." Clausewitz, on the other hand, emphasized "battle." . . . Here, his intention was focused on the bloody contest in the battle (just as Napoleon insisted on engaging the enemy before a decision could be brought about), rather than, as is the case of Sun Zi, on the situation prior to the battle.[22]

Finally, the Eastern commander is more flexible to what is by nature a fluid situation. As such, he is less susceptible to deception. The cost of sticking to one's initial plan through changing circumstances is often more casualties.

> Thirdly, Sun Zi emphasized adaptability to the changing situation and flexibility in command. . . . The law of successful operations is to avoid the enemy's strength and strike his weakness. . . . Hence, there are neither fixed postures nor constant tactics in warfare. . . . According to Sun Zi, numerical superiority alone does not count much, because "even if the enemy is numerically stronger, we can prevent him from fighting." . . . Clausewitz, however, stressed superiority of numbers and firmness in command.[23]

As a part of this emphasis on situation, the Asian commander more closely studies his potential adversary. His surveillance efforts extend to shadowing enemy forces and reconnoitering probable objectives from the inside. "When adequate information could not be gathered from reconnaissance outside the enemy camp, teams were prepared to go inside to gather the needed information."[24] As the Eastern commander prefers to keep these kinds of activities secret, firsthand accounts are extremely rare.

Once [in the Mekong Delta in 1965] I got into the middle of Cai Be [South Vietnamese Army] post where the district chief's office was. It was fifteen days prior to the attack and takeover of the post by our battalion. I was accompanied by two comrades armed with submachine guns to protect me in case my presence was discovered while I was nearing the post entrance. I was then wearing pants only and had in my belt a pair of pincers, a knife, and a grenade. At one hundred meters from the post I started crawling and quietly approached the post entrance with the two comrades following me. At twenty meters from the post, my comrades halted while I crawled on. At the post entrance there was a barbed-wire barricade on which hung two grenades. Behind the barricade stood a guard. I made my way between the barricade and the stakes holding up the barbed wire fence. I waited in the dark for the moment when the guard lit his cigarette. I passed two meters away from him and sneaked through the entrance. On that occasion I was unable to find out where the munitions depot was but I did discover the positions of two machine guns and the radio room. I got out at the back of the post by cutting my way through the barbed wire.[25]

> — Reconnaissance sapper
> 261st VC (Viet Cong) Battalion

Reconnaissance is not the only way in which an Eastern commander promotes secrecy. At the root of Oriental military thought is the combination of *zheng* (normal, direct force) and *qi* (extraordinary, indirect force).[26] The Asian engages his opponent with ordinary forces and beats him with extraordinary forces.[27] He can accomplish this in many ways. He may routinely use as a decoy the first troops to make contact with the enemy (e.g., use them as stationary base of fire while performing a secretive double envelopment). Or he may use conventional forces as a feint and then rely on nonconventional sappers or stormtrooper squads to destroy targets of strategic significance. For final victory, he may depend on special warfare methods. That Eastern armies carefully plan strategy is not to say that they centrally control all operations. When conducting mobile (or maneuver) war against a technologically superior opponent, they decentralize control.

The other aspect is the line of mobile warfare, the guerrilla character that is still needed at present in fighting both on a strategic and an operational scale, the inevitable fluidity of the base area, the flexibility and changeability of our construction plans in the base area, and the rejection of the untimely regularisation in building up the Red Army.[28]
— Mao Tse-tung

## More Tactical Options from Which to Choose

The Eastern infantry unit (unlike its Western counterpart) is equally adept at guerrilla, mobile, and positional warfare.[29] Its members can quickly shift between styles or temporarily exchange techniques.

The Oriental commander maximizes the fighting capacity of his unit by adjusting four variables: (1) formation or battle array, (2) advantageous position or military posture, (3) responsiveness to changing circumstances or flexibility, (4) and maneuvering the enemy or initiative (different from the Western connotation). He concentrates most on military posture — seeking out battlefield circumstances that closely mesh with the prerehearsed techniques of his subordinates. He must have discovered that a subordinate will demonstrate more flexibility and initiative after initially winning.

There are four indispensable factors in directing military operations, that is, battle array, military posture, flexibility and initiative. A thorough understanding of all will enable a general to bring the fighting capacity of his troops [in]to full play. The battle array that has to be drilled every day is like a sword that is worn by its user all the time. A sword can be used as an effective weapon only with the cooperation of its edge and handle, and a battle array can achieve victory only with the cooperation of its vanguard and rearguard units. . . . A precipitous military posture [like an ambush from a high place] is very powerful, just as an arrow shot from a bow can kill an enemy in a flash scores of metres away. . . . [To adapt to changing circumstances,] it is necessary to respond in different ways in military operations according to changes in the topography, environment and enemies. Taking the initiative is like using a weapon with a long handle. . . . Controlling the enemy instead of being controlled by him is just like attacking an opponent with a short sword by means of a long spear. . . . Of the four factors, military posture is the core. It is necessary to create a precipitous posture for complete victory by means of battle array, flexibility and initiative.[30]

— Sun Bin

To maintain the initiative, the Easterner relies heavily on deception. At the heart of deception is "varying the concentration and dispersion of forces."[31] To win each engagement, the Oriental infantry commander depends on forces that appear not to exist.

(Source: Corel Print House, *Plants*, Tree 11)

Sun Zi attached importance to the transformation of contradictions and stressed human beings' dynamic initiative in changing the disadvantaged into the advantaged, and the passive into the active.[32]

In terms of military strategy, there is the stratagem of complete victory by "subduing the enemy without fighting," and that of gaining initiative by maneuvering the enemy and avoiding being maneuvered. In terms of tactics, there is the stratagem of the complementation of weaknesses and strengths, and that of the combination of *zheng* [normal, direct force] and *qi* [extraordinary, indirect force].[33]

# The False Face
# and Art of Delay

- *In which ways does the Easterner show a false face?*
- *How does delaying his decision to act help him to win?*

(Source: *Handbook on Japanese Military Forces*, TM-E 30-480 (1944), U.S. War Dept., plate II; Corel Gallery, *Clipart* — Plants, 35C010)

## The 36 Strategies for Deception

The Oriental commander routinely wears a false face. It normally takes the form of a misleading posture or formation. It prevents his Western counterpart from acting on an accurate assessment of the situation — from initially controlling the "Boyd" or decision-making cycle. Central to this way of operating is the *yin-yang* antithesis from the *I Ching (Book of Changes)*. *Yin* and *yang* represent the endless series of opposites in nature. Things that are opposite can serve identical ends. For example, heaven and earth both nourish life. But *yin* is also paradoxically embedded in *yang.* Something's *yin* is like its reflection in a pool of water — embedded yet opposite. *Yin* is its unpredictable or secret aspect, and *yang,* its obvious or animate feature.

> A good war strategist either practices *yin* under the cover of *yang* or uses *yin* to supplement *yang.* The essence of the method lies in seizing [the] opportunity to make extraordinary moves.[1]
> — Jie Zi Bing Jing

The Oriental soldier routinely combines *shi* (bravery, strength, order, full stomach, leisure, numerical superiority, preparedness), with *xu* (cowardice, weakness, disorder, hunger, fatigue, numerical inferiority, unpreparedness); or *qi* (extraordinary) with *zheng* (normal).[2]  In wars with Western nations, the Asian's false face or misleading action is often what his opponent would do under similar circumstances, and his real face or intended maneuver is often the exact opposite.  Those who would go to war with an Easterner must learn to tell the difference.

When the Asian is ready to do what he has intended, it is almost always highly secretive and of strategic significance.  He may even have standby deceptions ready to counter events that would otherwise erode his cushion of surprise.  When his level of surprise becomes too low, he will simply pull back to try something else later.  To succeed against this skilled an opponent, one must study Oriental small-unit technique to discover which "sucker" moves routinely precede the hidden maneuvers.  Only then will a Western thinker be able to counter or preempt the Asian's hidden agenda.  A good place to start in this quest for Oriental technique is a compilation of 36 strategies for deception dating back to 200 B.C.  All modern versions descend from a single tattered copy found in mainland China in 1941.  Not until 1979 in Beijing was the book made available to the general public.

## The Imbedded Art of Delay

Of course, the Easterner does more than just mask intentions.  Through what has been called the "art of delay," he forces his adversary to make the first inappropriate move.  By practicing a little patience, the Easterner can not only discover the opposition plan, but often shift the momentum. Then, equally adept at the opposing styles of warfare (maneuver and attrition), he employs them like *yin* and *yang*. Instead of wildly pursuing the Westernized, high-speed version of the former, the Asian diligently works to acquire momentum. He knows that momentum must necessarily be the product of consecutive victories. He realizes that the frequency with which one wins in combat will largely depend on how often he can correctly surmise his foe's intentions, while concealing his own.

## Much of Maneuver War Gets Fought at the Mental Level

The 36 proverbs discuss more than just battlefield deceptions — they show how to undermine an opponent's sanity, integrity, and will to fight. As the goal of maneuver warfare is to demoralize rather than kill the enemy, the 36 stratagems should be of great interest to the aspiring maneuver warfighter. Literally interpreted, some are quite ruthless. Appendix A contains an in-depth discussion of aspects that obey the Geneva Conventions.

## The Most Powerful of the Deceptions

Of the 36 ruses, the last — running away — is probably the most important and least understood. When continuing to fight holds no strategic import, the Easterner will secretly withdraw. As subsequent chapters will demonstrate, he bases much of his battlefield methodology on this last, most ignominious, of the 36 proverbs.

> "Of the thirty-six strategies, running away is the best choice." This is a familiar remark, in literary works as well as in real life, from people who want to get around a situation that they are unable to cope with for the moment. The expression first appeared in the official *History of Southern Qi* about fifteen hundred years ago.[3]
> — Opening to the "Preface" of *The Wiles of War* Foreign Languages Press, Beijing

Tactical withdrawal and strategic retreat can not only assist one's war effort, but also save lives. By avoiding defeat, they pro-

**29**

vide for another chance to win. The preconceived notion that a victorious army advances and a defeated army withdraws can be misleading because it implies that moving backwards is always disgraceful. To intentionally limit one's freedom of movement on the battlefield makes no sense whatsoever.

Retreat can mean turning defeat into victory.[4]

To avoid combat with a powerful enemy, the whole army should retreat and wait for the right time to advance again. This is not inconsistent with normal military principles.[5]

Ancient Taoist sages invented the principle of "following the course of the times." It could be violated by either failing to take advantage of an opportunity or fighting a battle that could not be won.[6] The Asian commander knows that he must occasionally lose in combat. He sees dishonor only in refusing to acknowledge a defeat. He subsequently has no trouble learning from his mistakes.

Evade the enemy to preserve the troops. The army retreats: No blame.[7]

(Source: Courtesy of Stephan H. Verstappen, from *The Thirty-Six Strategies of Ancient China*, © 1999 by Stefan H. Verstappen)

## Stratagems When in a Superior Position
Openly Cross an Open Area
Seize Something Valuable the Enemy Has Left Unguarded
Use the Enemy against Himself
Make the Enemy Come to You
Capitalize on a Natural Disaster
Pretend to Attack from One Side and Attack from Another

## Stratagems for Confrontation
Make Something out of Nothing
Attack from One Side and Switch to Another
Give Murphy's Law Time to Work on the Enemy
Mask a Sinister Intention with a Good Impression
Sacrifice Minor Concerns for the Sake of the Overall Mission
Seize the Chance to Increase the Odds

## Stratagems for Attack
Make a Feint to Discover the Foe's Intentions
Steal the Enemy's Source of Strength
Draw the Enemy Away from His Refuge
Give a Retreating Adversary Room
Discover a Foe's Intentions by Offering Him Something of Value
Damage the Enemy's Method of Control

## Stratagems for Confused Situations
Erode the Enemy's Source of Strength
Create Chaos to Make an Opponent Easier to Beat
Leave Behind a Small Force to Slow and Deceive Pursuers
Encircle the Foe but Let Him Think He Has a Way Out
Concentrate on the Nearest Opposition
Borrow from the Enemy the Instrument of His Own Destruction

## Stratagems for Gaining Ground
Attack the Enemy's Habits to Undermine His Foundation
Make an Example of Someone to Deter the Enemy
Feign Lack of Military Ability
Lure the Foe into Poor Terrain and Cut Off His Escape Route
Mislead the Enemy with False Information
Secretly Occupy an Enemy-Controlled Area

## Stratagems for Desperate Situations
Lull the Enemy to Sleep with Something Beautiful
Reveal a Weakness to Cause the Enemy to Suspect a Trap
Attack the Enemy's Cohesion from Within
Feign Injury to Yourself
Combine Stratagems
Refuse to Fight

### Figure 3.1: Deception Recap

## The Sister Stratagems

"Running away" is the last of six ways to handle desperate situations. All have a common thread — using gestures of weakness, enticements of gain, and apparent defeats to lure an arrogant opponent into unfavorable circumstances. The fourth of the six is, "Feign injury to oneself (or one's unit)." The first is, "Lull the enemy to sleep with something beautiful." To hide his individual and small-unit skill, the Asian may on occasion admit to a defeat he didn't suffer or casualties he didn't take.

[S]ubmit flattering reports to blunt him [the enemy].[8]
— Liu Tao, *Six Strategies*

## What May Have Actually Happened

What occurred in history does not change, but one's perception of it does — as he comes to better understand his opponent. There is no telling how many fake or evacuated enemy emplacements may have been successfully bombed and assaulted by U.S. forces. Opposition losses may have been unintentionally exaggerated throughout much of the 20th century. With the acknowledgment of too many enemy casualties estimated may come the realization of too many American lives lost. To suffer fewer casualties in war, the United States must look to the east.

The supreme art of war is to subdue the enemy without fighting.[9]
— Sun Tzu

# The Hidden
# Agenda

- *How do Easterners destroy targets of strategic import?*
- *Which "false face" precedes each actual maneuver?*

(Source: *Handbook on Japanese Military Forces*, TM-E 30-480 (1944), U.S. War Dept., plate II; Corel Gallery, *Clipart* — Plants, 35A024)

## How the Eastern Battle is Fought

The Oriental infantry commander tries hard to function within the confines of reality. He is a student of every aspect of the combat situation, and seldom asks his subordinates to exceed their capabilities. His thought processes have been heavily influenced by Taoism — an ancient philosophy promoting self-realization through patience, simplicity, and harmony with nature.[1]

In my opinion, the importance of understanding the Taoist element of *The Art of War* can hardly be exaggerated. Not only is this classic of strategy permeated with the ideas of great Taoist works such as the *I Ching (The Book of Changes)* and the *Tao-te Ching (The Way and Its Power),* but it reveals the fundamentals of Taoism as the ultimate source of all the traditional Chinese martial arts.[2]

— Thomas Cleary in "Preface" to *The Art of War*

In battle, the Easterner commander adapts to his environment. Whenever possible, he fights at a time and place that meshes well with the abilities of his unit. To further increase his odds of victory, he studies what is about to become a fluid situation. Before the battle, he roots out obscure circumstances. During the battle, he allows his people to respond to altered circumstances. In other words, he bases his approach on planning and flexibility. Through diligent reconnaissance and rehearsal, he will try to win the battle before it is fought. His small units will have practiced their techniques so often as to be able to wield them like a sword. He will have concentrated on the same four factors emphasized by Sun Bin[3] — battle array, military posture, flexibility and initiative. Then, to thoroughly baffle and further feel out his opponent, he will employ a "false face" and "temporary delay."

Fighting is not only a battle of courage, but also of wits. Hence, creating false formations and illusions enables one to be camouflaged. This will confuse the enemy.[4]
— *100 Strategies of War*

The Easterner's approach to combat does little more than obey common sense at the squad level. Yet, it is quite different from that in the West. The American, British, or French commander will "march to the sound of the guns" and then look for the wherewithal with which to succeed. He may cursorily scan the battlefield from afar but then pins his unit's chances of victory almost entirely on his own tactical-decision-making and leadership ability.

## Ancient Battle Arrays or Formations

When studying Eastern formations, one must remember how the Asian commander uses his ordinary and extraordinary forces. Sometimes, he sends large units to alternately grind down his opponent from the front. Other times, his main force demonstrates to the front while tiny, covert elements work around back. Most of the time, he uses the threat of frontal assault to keep his foe from noticing how much damage the sapper teams and stormtrooper squads are doing to his strategic assets. The Asian will almost never reveal his true battle array in its entirety.

Below are described the ancient formations from which the modern equivalents derive.

(Source: Courtesy of Stephan H. Verstappen, from *The Thirty-Six Strategies of Ancient China*, © 1999 by Stefan H. Verstappen)

*The Cord Battle Array*

The "Cord" is like the 19th-century Zulu "Horns of the Buffalo" attack formation. The quarry is preoccupied with a slow-to-develop frontal attack while being secretly double enveloped.

> The Cord Battle Array is used to attack and annihilate the enemies. It . . . is suitable for attacking weak enemies with strong forces.[5]
> — Sun Bin

*The Obstructing and Blocking Array*

"The Obstructing and Blocking Array is used to wear out the enemies."[6]  Obstructing means restraining, and blocking means intercepting.  This formation is shaped like a concentric nest of the

letter "C." Why the circle openings face the enemy is not totally clear. Perhaps they act as a firesack — large enough to draw in the quarry, but small enough to cover the void by fire.

Although partial, the concentric circles give this formation a defense-in-depth capability — something seldom accomplished in the West. Its modern equivalent may be the "Parallel-Line Defense." It will be discussed in Chapter 8.

*The Cloud Battle Array*

Unique to the Orient is the "Cloud" maneuver. It allows the Eastern commander to disperse or concentrate his forces as the situation dictates. To accomplish this quickly, those forces need communications and rehearsal. Among its modern applications may be the "Starburst" defense discussed in Chapter 8.

> The Cloud Battle Array is used when both sides are shooting at each other with bows and arrows [or indirect fire]. Such a battle array is flexible in making changes and gathers and disperses troops quickly.[7]
> — Sun Bin

> Before engaging in [any] battle one must first train the soldiers. . . . The troops must learn the strategy of how to assemble and disperse. Follow the commands of attack, lying still, advance and retreat.[8]

*The Yingwei Battle Array*

"The Yingwei Battle Array is for besieging the enemies."[9] Its modern equivalent may be the "Parallel Urban Thrusts" described in Chapter 11.

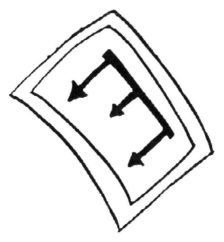

*The Hesui Battle Array*

"The Hesui Battle Array is for wiping out the enemy vanguard units."[10] Shaped like a sawtooth, its modern equivalent will be discussed in Chapter 6.

*The Three-Shaped Battle Array*

This formation is for attacking defeated enemies.[11]  Shaped like the number "3," its opening faces the foe. The enemy unit would be split, and its halves encircled.

*The Cone-Shaped Battle Array*

"[The] Cone-Shaped Formation is for breaking through the strong defence lines and the crack troops of the enemies."[12]  Its modern equivalent must be similar to the U.S. "Wedge."

*The Flying-Geese Battle Array*

"[The] Flying-Geese Shape Formation is to raid the enemy flanks."[13] The opening faces the enemy.

*The Closing-V or Haichi Shiki Battle Array*

The actual vintage of the "Closing-V" or *Haichi Shiki* maneuver is not known. Again, the opening faces the foe. For counterattacking a pursuer, its sequel was encountered in Korea

> [T]he Chinese Army . . . planned attacks to strike from the rear, cutting escape and supply routes and then sending in frontal waves. The basic battle tactic was the *Haichi Shiki*, a V formation into which they allowed opposing troops to move. The Chinese then would close the side of the V while another force moved below the mouth to stop any attempts at escape and to block relief columns.[14]

## The Circle Formation

As Oriental units do not normally employ perimeter defenses, the "Circle Formation" deserves special note. "If attacked from four sides, form a circle and fight from all sides."[15]

## The Split-Unit Formation

While the "Split-Unit Formation" was initially designed to counter horsemen, it might also discourage a larger infantry force. It may the forerunner of the "Wagon Wheel Defense" described at the end of Chapter 8.

> [I]f attacked from two sides, split the army to harass the enemy from behind. . . . If attacked from one side, fight the enemy from both sides to cover one another.[16]

On 18 August 1968, in what was to become the final battle of the Second Tet Offensive, an NVA (North Vietnamese Army) platoon got caught by a U.S. Marine cordon operation at the village of "La Thap (1)" northwest of An Hoa. Under attack from two sides, the enemy commander sent a single soldier around to snipe the U.S. maneuver force commander from the rear.[17]

Several months later and 1500 meters to the south, another North Vietnamese unit got caught between the same two Marine companies near the village of "Chau Phong (1)." That time, an enemy machinegunner stayed behind to pin down the U.S. units while his comrades escaped through an unoccupied gully between them.[18]

## The Military Posture or Planned Situation

The Easterner carefully chooses when and where to fight. Ambushing one's opponent from a precipitous place would be an example of good military posture.

## The Flexibility to React to Unforeseen Circumstances

The Asian can readily adapt to what normally constitutes a fluid situation in combat. He has several ways in which to respond to changes in topography, environment, and enemy intentions. He may actually enjoy more leeway to act upon what he sees than his British, French, or American counterpart.

> When fighting, one should assess the actual situation to adopt suitable strategies. In the midst of fighting, be resourceful and be adept in seizing opportunities. This is a principle military strategists have always abided by since the olden times.[19]
> — "Deployment of Foot Soldiers"
> *100 Strategies of War*

> Look out for the enemy's weakness, vary the strategy according to the changing circumstances and avoid engaging a strong, well-prepared enemy.[20]
> — "Fighting a Well-Prepared Enemy"
> *100 Strategies of War*

Rule: To achieve victory, one has to adapt to the changed situation accordingly.[21]
— "Being Flexible Enough . . ."
*100 Strategies of War*

Of course, to quickly identify and then react to changing circumstances, one must harness the perceptions and common sense of the lower-ranking members his unit.

Arrogate all authority to oneself but pool the wisdom of the masses. . . .
Victory or defeat depends on the judicious employment of people. A shrewd general must be able to employ his people judiciously, and take full advantage of their talents.[22]
— "The Wisdom of Military Leadership"
*Six Strategies for War*

## Initiative

In the East, the word "initiative" has a broader connotation than in the West. The Asian displays initiative every time he controls his adversary or avoids being controlled by him. To Mao Tse-tung, the term meant "freedom of action." For Mao, initiative could sometimes be exercised by running away.[23] Of course, the Easterner has many ways to display initiative against a stronger opponent.

a) *To lure the enemy,* by diverse means, to make him fall into a trap prepared by us. . . .

b) *To attack the enemy's exposed and weak points* in order to make it impossible for him to defend himself. . . .

c) *To move skillfully,* advancing rapidly towards the enemy rear, attacking important points, with a view to obliging the enemy troops to regroup, thus upsetting their predetermined battle plan; to strike the enemy on his flank and from behind, rather than throwing ourselves directly at his main front.

d) To clearly know the enemy's situation in order to . . . *concentrate our regular troops rapidly and move our reserve forces swiftly to the required areas to act in good time.* . . .

e) To centralize the leadership in the hands of a higher
command. But the local command *can and must act
according to the situation at the front,* and must fight
according to their own initiative in order not to miss
good opportunities or to fall into a state of passivity.[24]
— Truong Chinh, *Primer for Revolt*

"To keep the initiative is the essential principle of [Eastern]
tactics in general, and of guerrilla and mobile [maneuver] warfare
in particular."[25]

Variability is the essence of military strategy. Birds gather
and disperse in infinite variations. . . . Likewise, military
stratagems have to vary according to battleground
conditions.[26]
— *Six Strategies for War*

## The Combined Power of Several Warfare Styles

The Oriental commander, unlike his Western counterpart, has
the uncanny ability to shift rapidly between warfare styles.
Unfortunately, he can do more than just alternate the *yin* of
maneuver with the *yang* of attrition. By 1945, Mao Tse-tung was
talking of three styles — guerrilla, mobile, and positional — and of
the transitions between them.[27] In protracted war, he identified
mobile as his primary and guerrilla as his secondary.[28]

In fact, an Oriental army can conduct guerrilla and mobile
warfare at the same time. It uses the former to help coordinate the
campaigns of the latter. As will be discussed in Chapter 10, North
Vietnamese maneuver forces relied on Viet Cong for much of their
security and reconnaissance.

The great role of strategic coordination played by
guerrilla warfare should not be overlooked. The leaders of
guerrilla armies and regular forces should clearly grasp its
significance.

Moreover, guerrilla warfare also performs the function
of coordination in campaigns.[29]
— Mao Tse-tung

(Source: FM 21-76 (1957), p. 54)

Of course, key to coordinating guerrilla and conventional warfare from the same headquarters is a willingness to decentralize control over all operations. Just to survive, guerrillas must disperse, concentrate, or shift as the situation dictates. To retain any semblance of initiative, conventional soldiers must also make many of their own decisions. To keep from becoming too slow or predictable, they too need simple instructions and non-uniform methods.

> [G]uerrilla units are armed bodies on a lower level characterized by dispersed operations. . . . If we [the Chinese] attempt to apply the [old] method of directing regular warfare to guerrilla warfare, the high degree of elasticity of guerrilla warfare will inevitably be restricted and its vitality sapped. . . .
> . . . [I]t demands a centralised command in strategy and a decentralised command in campaigns and battles.[30]
> — Mao Tse-tung

Most disturbingly, the Oriental armies may have discovered how to incorporate guerrilla tactics into regular warfare — to further refine the German techniques generally considered to be state-of-the-art. In the third stage of Mao's protracted war against the Japanese, guerrilla warfare was allowed to evolve into mobile war.[31]

What was learned in the process — in the areas of tactical technique, operational control, and training — may have been significant. The Asians may have learned how to conduct conventional warfare in an unconventional manner.

> [I]t becomes possible for the guerrilla units to go through the necessary process of steeling and to change gradually into regular armies; consequently, with their mode of operations gradually transformed into that of the regular armies, guerrilla warfare will develop into mobile warfare.[32]
> — Mao Tse-tung

> Moreover, a regular force, when dispersed, can conduct guerrilla warfare, and, when reassembled, can resume mobile warfare, just as the Eighth Route Army has been doing.[33]
> — Mao Tse-tung

In Korea, the Chinese managed to invade a country without ever appearing to invade. To fight U.S. forces to a standstill, they needed no heavy weapons and very few supplies. Like the Japanese before them, they relied almost entirely on surprise.

> The Chinese army was largely a guerrilla force that was extremely mobile because it did not employ any heavy weapons. In fact, its soldiers — whose largest weapons were light mortars — carried everything on their backs.[34]

Then 30 years later in Vietnam, it became impossible to ignore this evolution in conventional warfare. There, the foe needed no tanks, planes, or dependable supply lines to defeat the richest and most technologically advanced nation on earth. For the first time, enemy maneuver forces dispersed during their approach marches — sending subordinate elements in along different routes. After those elements secretly rendezvoused near the objective, their parent unit would attack. Then the parent would subdivide one last time — sending its parts back out along new routes.

When North Vietnam — with its expeditionary forces off fighting the Khmer Rouge in Cambodia — inflicted 20,000 casualties on 30 Chinese divisions in a 17-day border war in February of 1979, it established which was the best of the best.[35]

# Part Two

## The Differences in Tactical Technique

He who knows when he can fight and when he cannot
will be victorious. — Sun Tzu

# Ghost Patrols and Chance Contact

- *How can an Eastern patrol disappear from view?*
- *What does its point security element do differently?*

(Source: Courtesy of DeAgostini UK Ltd., from *The Armed Forces of World War II*, © 1981 by Orbis Publishing Ltd.; Corel Gallery, *Clipart — Plants, 35A023*)

## The Eastern Patrol Leaves Little Trace

Not much has been written about Asian patrols, for they are seldom seen on the move. There may be several explanations for their invisibility: (1) skilled point men, (2) expert camouflage, (3) willingness to break contact, and (4) preference for crawling.

In essence, Oriental patrols generally spot their Western counterparts first. Exceptions may have occurred on Guadalcanal, where the U.S. Marine Raiders had been trained like the Communist Chinese. The patrol that sees another first almost always has better point men.

## The Crawling Point

One can readily imagine the ideal point man. He would have above-average senses and infinite patience. He would be able to move unseen through every type of terrain. That takes one of two things — an expertise in stalking or an affinity for crawling. An upright human does not look for one of his species on the ground. What a Marine NCO witnessed one day on Guadalcanal may reveal a key element in the Oriental patrolling technique.

> Here's the way the Japs patrol. I was out on the bank of a river with another man. We were observing and were carefully camouflaged. We heard a little sound and then saw two Japs crawl by about 7 feet away from us. These Japs were unarmed. We started to shoot them, but did not do so as we remembered our mission. Then, 15 yards later came 8 armed Japs. They were walking slowly and carefully.

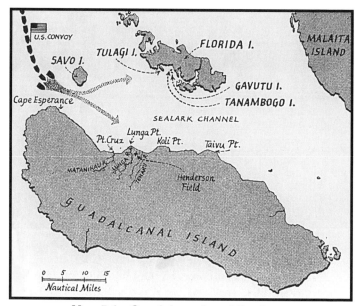

**Map 5.1: Guadalcanal and Environs**

(Source: Courtesy of HarperCollins, from *Guadalcanal*, by Irving Werstein, copyright © 1963 by Irving Werstein)

We did not shoot as our mission was to gain information. When I got back, we had a lot of discussion as to why the two Japs in front were not armed. . . . I believe they were the point of the patrol and were unarmed so they could crawl better.[1]
— Sergeant O.J. Marion
FMFRP 12-110

## The Occasional "Shoot and Run" Deception

Unlike the American, French, or British soldier, the Oriental soldier sees no shame in moving backwards when it serves no purpose to stay and fight. On 6 November 1942, Evans Carlson's 2nd Raider Battalion took a month-long patrol behind Japanese lines on Guadalcanal. The first few Japanese they encountered took a few shots at their native guides and ran away.[2]

While considered less than manly in the West, running away has always been tactically valid in the East. It helps a valued resource to live to fight another day. It also helps him to bait a trap. Many a U.S. unit has been decimated trying to chase down one or two potential prisoners. Those units invariably make the mistake of entering an area with too little cover. Ten-foot-tall elephant grass can do little to impede grazing machinegun fire.

## The More Common Covert Action

When an Asian point man sees an enemy patrol approaching along the same trail he occupies, he signals for a hasty ambush — for his comrades to individually turn and move a short distance off the trail. Evidence of this maneuver — matted grass at regular intervals — has been observed along the trails of another battlefield that lies hundreds of miles to the northwest.[3]

When a Japanese patrol suspected an American presence atop Mount Austen on Guadalcanal, it followed that same procedure. While not very aggressive, this technique does provide its practitioners with more cover.

A few minutes after Lieutenant Jack Miller gained the "hub" [confluence of fingers atop Mount Austen], a strong enemy

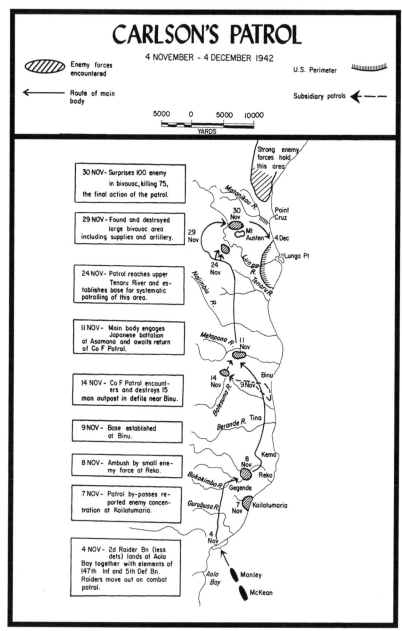

# CARLSON'S PATROL

### 4 NOVEMBER - 4 DECEMBER 1942

Enemy forces encountered

Route of main body

U.S. Perimeter

Subsidiary patrols

5000    0    5000    10000

YARDS

Strong enemy forces hold this area

30 NOV - Surprises 100 enemy in bivouac, killing 75, the final action of the patrol.

29 NOV - Found and destroyed large bivouac area including supplies and artillery.

24 NOV - Patrol reaches upper Tenaru River and establishes base for systematic patrolling of this area.

11 NOV - Main body engages Japanese battalion at Asamana and awaits return of Co F Patrol.

14 NOV - Co F Patrol encounters and destroys 15 man outpost in defile near Binu.

9 NOV - Base established at Binu.

8 NOV - Ambush by small enemy force at Reko.

7 NOV - Patrol by-passes reported enemy concentration at Koilotumaria.

4 NOV - 2d Raider Bn (less dets) lands at Aola Bay together with elements of 147th Inf and 5th Def Bn. Raiders move out on combat patrol.

Matanikau R.
30 Nov
Point Cruz
29 Nov
Mt Austen
4 Dec
Lunga Pt
Lunga R.
24 Nov
Nalimbiu R.
Tenaru R.
Metapona R.
11 Nov
Binu
14 Nov
9 Nov
Bokesuna R.
Berande R.
Tina
8 Nov
Kema
Reko
Bokokimbo R.
Gegende
Gurubusa R.
7 Nov
Koilotumaria
4 Nov
Aola Bay
Manley
McKean

## Map 5.2: Marine Raider Patrol

(Source: *U.S. Marine Corps Special Units of World War II*, Hist. & Museums Div., HQMC, p. 20)

combat patrol was observed approaching along one of the ridges to the east. As it neared the crest, it apparently became suspicious and deployed.[4]

[T]heir leader was war-wise, and he suddenly sensed that something was very much wrong. He shouted, and all the Japanese dropped off the trail and into the brush alongside.[5]

## The Reconnaissance by Fire and Subsequent Maneuver

While certainly aware of the effect on surprise of small-arms fire, the Japanese patrols on Guadalcanal appeared more than willing to perform reconnaissance by fire.

They [the same Japanese patrol on Mount Austen] opened fire.[6]

Drawing the Marines' fire helped the Japanese in two ways: (1) to prematurely trigger any ambush, and (2) to pinpoint their foe's location. To save ammunition, the Nipponese would occasionally snap bamboo sticks together.[7] Often, their bullets were not well directed.

The Japanese fire is not always aimed. It is harassing fire and scares recruits.[8]
— One of Chesty Puller's NCOs
FMFRP 12-110

But the shooting may have helped the enemy in other ways as well: (1) to summon other Japanese patrols, (2) to capture the opponent's attention, and (3) to cover the sound of movement.

[T]hen the Japanese [on Mount Austen] decided to outflank the Americans.[9]

## The Encirclement

The Japanese enlisted men did not run away or jump off the

trail because they were afraid, they did so because they had been allowed to obey their common sense. What the Western soldier must come to understand is that surprise-oriented armies prefer not to leave any witnesses to their passage. If an Oriental force discovers a smaller opponent in a meeting engagement, it will try to annihilate that opponent. Once it has established a base of fire, it will send out separate maneuver elements to the left and right flanks. It happened that way on Mount Austen, just as it had several days before.

> Carlson succeeded in executing a double envelopment. This was accomplished only after a similar attempt on the part of the enemy had been circumvented.[10]
> — *The Guadalcanal Campaign*
> History Division, HQMC

> [Earlier,] Company E [2nd Raider Battalion] under Captain Washburn [had] moved up the west bank of the Metapona [River], and just as they reached Asamana, surprised two companies of Japanese who were crossing the river from east to west. . . . As the Raiders came up and fired into the Japanese, enemy outposts opened up from good positions. A machinegun in the fork of a banyan tree was particularly effective and succeeded in pinning the Raiders down. . . . Then the Japanese forces began a flanking maneuver to get behind Captain Washburn's company. He withdrew to reorganize. . . .
> Company E returned, however with a fresh platoon . . . in the lead. This time they knocked out the machinegun in the banyan tree, but . . . the Japanese again began a flanking movement from two directions. As one force tried to work to the west of Washburn's company, six machineguns opened up on the Marines from the opposite bank.[11]

## The Tightening Noose

Once the encirclement has been completed, Oriental soldiers will close in for the kill. Their fondness for crawling suggests the method. If all were to crawl simultaneously toward their besieged victims, shoot only at those who weren't watching, crawl forward before shooting again, and take turns moving and firing, could they not systematically destroy their opposition with very little risk to themselves?

Nothing like this was recorded on Guadalcanal, but it was many years later on the Asian mainland.[12]

## The Alternative Maneuver

If the terrain will not conceal a double envelopment, Oriental soldiers will sometimes frontally assault. But this will be an assault quite different from those in the West. The assault troops will be visible only while moving, and not all will move at once. On Guadalcanal, Carlson's Marines witnessed a skirmish line of bushes.[13]

> On the morning of the second day (November 13) [at Assamana] it was discovered that . . . the remnants of the force hit by Captain Washburn lay in the woods to the west and north. . . .
> At one point a Marine lookout reported the remarkable news that a forest was moving toward the Raiders' position. Through field glasses Colonel Carlson saw that in-

(Source: FM 21-76 (1957), p. 56)

deed Birnam Wood was moving. . . . A company of Japanese, in mass formation, was marching across a field of grass wearing cloaks of foliage and twigs.[14]

Japanese infantrymen could blend in with more than one kind of terrain — from jungle to savannah. Because they moved in short, unobtrusive rushes, they were hard to see long enough to shoot.

The Japanese sew grass and leaves to their shirts and hats.[15]
> — One of Chesty Puller's NCOs
> FMFRP 12-110

## The Assault

During any ground assault, the Japanese preferred grenades and bayonets to small arms. They wanted their projection of force to sound to the majority of defenders like a mortar attack. They experienced less opposition that way. Even the Marines in their path had trouble telling what was going on.

## More Trouble on the Way

Asians will intentionally send more than one patrol into the same area. The job of some may be to shadow others. Whenever one makes contact, others rally to its support and arrive from unexpected directions. As Marine Raider Evans Carlson had learned this technique from Mao Tse-tung before WWII, any Oriental unit might use it.

When one patrol found the enemy, all the units would be brought to that point to make a concentrated attack. This was another bit of borrowing from the Chinese communist system of guerrilla warfare; Mao Zedong's concept was that small enemy units should be cut off, enveloped, and destroyed.[16]

# The Obscure, Rocky- Ground Defense

● *How are Eastern bunkers hidden on barren ground?*

● *In what way do they constitute a trap?*

(Source: Courtesy of Cassell PLC, from *World Army Uniforms since 1939*, © 1975, 1980, 1981, 1983 by Blandford Press Ltd.; Corel Gallery, *Clipart* — Plants, 35A036)

## The Eastern Defender

By 1918, the largest nations in Eastern Europe were practicing something similar to the Oriental style of war. The German version came to be known as maneuver or "common-sense" warfare. What this "merging of philosophies" has meant to U.S. forces is inescapable. Throughout most of the 20th century, they have had to deal with defensive tactics quite different from their own — tactics specifically designed to thwart their edge in firepower.

Stopping a more heavily armed attacker takes planning and flexibility. One must carefully choose when and where to fight, surprise his opponent, and then quickly withdraw. Maneuver or "common-sense" warfare offers several ways in which to do this: (1) ambushes in series, (2) strongpoints to channel tanks into preplanned killing zones, (3) reverse-slope defense, (4) soft or elastic defense in depth, and (5) not physically occupying the ground to be defended. All are intended to lay the groundwork for an eventual counterattack.

But the maneuver warfighter works just as hard to demoralize his opponent as to physically defeat him. He capitalizes on the old adage that "truth is the first casualty of war" by initially disguising his maneuver and then encouraging his opponent to ignore it. On defense, he accomplishes the latter by inflicting serious casualties and then mysteriously disappearing. With little to show for his sacrifice, the prideful attacker becomes tempted to inflate the body count. By repeatedly opening the door to what isn't true, he becomes less interested in the uniqueness of prebattle circumstances. His tactical decisions begin to follow a pattern. He may even start to believe that he has won every engagement. At that point, the maneuver warfare defender has helped himself immeasurably. The opponent who never gets defeated has little reason to improve between battles.

The maneuver war defender also encourages his adversary to act first on what he has been allowed to see. Then, with speed, stealth, and deception, he launches his own attempt to influence the outcome of the war. This attempt may not be readily apparent to the adversary. Often, it can only be surmised later by students of the alternative way of fighting. Luckily, most of America's 20th-century adversaries have exhibited a similar *modus operandi*.

The best example of what a maneuver war defender can accomplish occurred on a tiny volcanic island 650 miles south of Tokyo in February and March of 1945. The mere mention of "Iwo Jima" brings tears to the eyes of those who were there. Let no one doubt that those brave young Americans did a fine job overcoming what historians have come to regard as a defensive masterpiece.

## Escaping the Pre-Invasion Bombardment

Hotly debated before every U.S. landing has been the extent to which coastal areas (and local populace) should be subjected to preparatory fire. The shells and bombs have made lots of noise over the years but have had less effect on the enemy.

Iwo's defenses were nearly perfect — (1) elsewhere than expected, (2) invisible to the naked eye, and (3) impervious to bombardment. In fact, they were almost entirely below ground. As the Japanese contested the beaches with long-range fire, much of the pre-invasion shelling landed on unoccupied ground. Even the direct hits did little good.

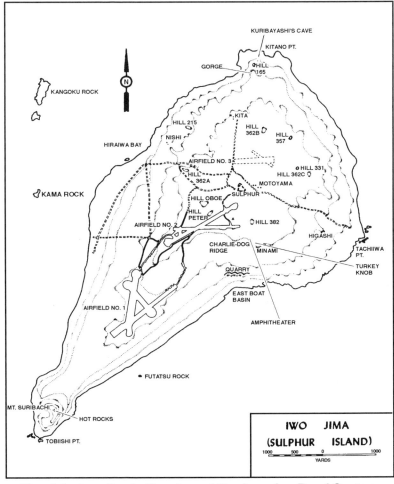

## Map 6.1: Iwo Jima's Terrain and Pre-Invasion Road System

(Source: *Closing In: Marines in the Seizure of Iwo Jima*, Hist. & Museums Div., HQMC, p. 5)

But the shrewdness of General Kuribayashi, and the way he had built and placed his fortifications, made shelling and bombing largely useless. The positions were mas-

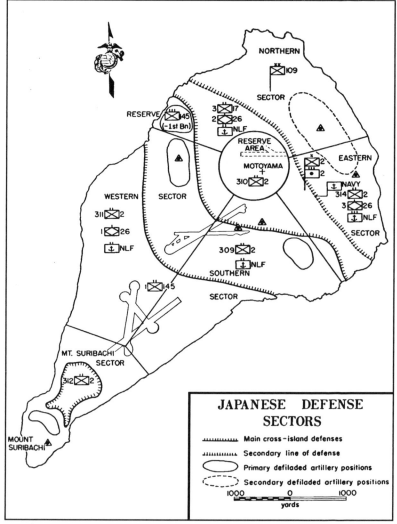

**Map 6.2: Iwo Jima's Defense Lines**

(Source: *Closing In: Marines in the Seizure of Iwo Jima*, Hist. & Museums Div., HQMC, p. 8)

terpieces of concealment and construction: walls of many were more than three feet of steel-reinforced concrete and impossible to spot from the natural terrain upheavals that camouflaged them. Some measured ten by twenty feet, had five-foot ceilings, and were made up of three separate rooms with twelve-inch concrete walls, the areas connected by narrow crawlways.

Apertures for cannon were virtually invisible, jutting from rubble of [the] volcanic boulders. Atop many of the installations were pillboxes . . . all but impregnable fortresses that could take a direct hit from naval shellfire or a bomb . . . without suffering much damage. Even when Marine demolitions men could fling a satchel charge into a gun vent, the blast would have little or no effect on the other two rooms.[1]

## Springing the Trap from Standoff Distance

Every aspect of a maneuver warfare defense — whether large or small — is a trap. Because this alternative way of fighting has the power to shift the momentum, it is often used as an offensive tool.

When the thousands of Marines initially landed on Iwo Jima, there was little shooting. They took cover behind the terrace of black sand beyond the beach. Then long-range fire from the flanks (Suribachi and the Quarry) took a terrible toll.

## The Near-Ambush Deception

A close-quarter defense can shift the momentum in battle much as a goal-line stand can in football. Past adversaries have taken full advantage of a standard U.S. procedure — frontally assaulting any ambush within 50 yards. So easy is it to create the impression of close ambush that any number of ways have been devised to lure additional victims. Having baited the trap, the maneuver warfare defender carefully hides. To escape U.S. supporting arms, he holds his fire until his quarry is almost upon him.

This alternative way of fighting worked well in the jumbled terrain of Iwo Jima's northern highlands where 25-yard fields of fire were not uncommon. Routinely, Marines took fire simultaneously from front, flanks, and rear. Several explanations are possible — e.g., firesacks or bypassed positions. But the effect on the Marines was always the same — as if they had just been ambushed from several different directions at once.

Ninety minutes after landing [on D-Day], the battalions of Lieutenant Colonel John A. Butler and Major John W. Antonelli had clawed up the slope to the eastern edge [of Airstrip Number One]. Resistance was light until 10:45 [A.M.] when, as Antonelli put it, "crap hit the fan in copious quantities." Moments before, they were catching only small arms fire. Now, when Kuribayashi pulled the plug, they were in the middle of the ambush and taking a deadly fusillade of machinegun bursts and mortar fire from concealed bypassed positions in the hummocks.[2]

## The Unseen Presence

Some of America's adversaries have fortified so extensively below ground as to give *terra firma* more than one level. They have, in effect, turned rural terrain into urban terrain and, by so doing, acquired a decided edge over their more powerful opponent.

Most of the Japanese bodies on Iwo Jima were never recovered. To even begin to imagine what it must have been like to be there, one must come to grips with several paradoxes. First and foremost was the absence of a visible opponent. Though practically devoid of vegetation, Iwo's volcanic terrain appeared uninhabited during the fiercest fighting. Seeing one's buddies shot and then having no one to shoot back at was hard on morale.

We hardly ever saw an enemy. The Japanese had every inch of ground covered by fire.[3]
— Navaho Marine veteran
"Japanese Codetalkers" on History Channel

Seldom on Iwo, from D-day until the battle was over [36 days later], did you see the enemy — just the sights and sounds of deadly fire from his weapons.[4]
— Marine infantryman on Iwo Jima

How did the Japanese manage to blend in so well with the barren landscape during prolonged, close-range combat? Basically, they did everything imaginable — from using flashless and smokeless powder to firing from the shadows behind narrow bunker apertures (an urban-warfare technique). The front of each machinegun fort or "strongpoint" was swept by the fire of others behind it. By focusing on the immediate objective, Marines had difficulty locating the source of additional fire. To make matters worse, the occupants of these well-dispersed strongpoints and subsidiary "outposts" took turns shooting — keeping the Leathernecks guessing as to where the bullets were coming from. But key to the defender's unprecedented disappearing act was the extent to which he camouflaged each position. When visible, machinegun apertures consisted of narrow slits, barely a foot wide and at ground level.[5] Before spewing death, an open aperture may have looked like a natural crevice below a rock outcropping. Unfortunately for the Marines, many of these apertures had camouflaged, removable covers:

One of the wounded Marines we received aboard when the hospital ships were overtaxed, declared the Japanese had

**Figure 6.1: Typical Japanese Bunker**
(Source: *TM-E 30-480* (1944), p. 159)

**63**

camouflaged trapdoors all over the face of the cliff and that they would open a door, pour out a murderous volley, and close the door again, leaving the attackers staring at an apparently blank area of mountainside.[6]

## Withdrawal

On Iwo, the Marines encountered defenses quite different from their own. Instead of continuous, barbed-wire-protected trenchlines or strings of fighting holes, they found what the Germans had used late in WWI — a matrix of semi-independent machinegun forts, all of which were mutually supporting, and any one of which could be abandoned at the discretion of its NCO in charge.[7]  Contrary to what they had been told, *bushido* only occasionally interrupted this

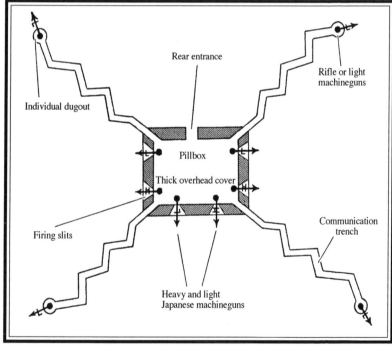

**Figure 6.2: Strongpoint Complex**
(Source: *TM-E 30-480* (1944), p. 160)

**Figure 6.3: Pillbox Interior**
(Source: *TM-E 30-480* (1944), p. 161)

pattern of tactical withdrawal. Documentary film footage in the movie *Sands of Iwo Jima* clearly shows a small group of Japanese moving diagonally backwards from the main bunker on Tarawa.[8]

However, there were some important differences between German and Japanese method. The WWI German machinegunners had 50-foot-deep "dugouts" in which to weather the shelling and trenches through which to fall back. The WWII Japanese machinegunners had nearby caves or pillboxes in which to escape bombardment and tunnels through which to pull back. Common throughout the earlier island campaigns of WWII had been a blockhouse connected by tunnel or trench to a series of automatic-weapon emplacements. The variation that occurred on Iwo Jima was a semicircle of automatic-weapon positions at each end of the tunnel through a ridgeline. Rear-strongpoint occupants could then conduct a reverse-slope defense or remain hidden to shoot Marines in the back as they assaulted the next objective. Under either scenario, the Nipponese had easy access to bombardment protection and an avenue of egress. In other words, the reverse slopes were as strongly fortified as the forward slopes.[9] As these strongpoints (on either slope) were carefully concealed and routinely mortared from behind (to brush off intruders),[10] attacking Marines had difficulty establishing their precise location. In a linear defense, mortar rounds would be called on machinegun "dead space." They gave the impression of unoccupied ground.

The Japanese called being on defense "retreat combat."[11] On Iwo Jima, they executed a "soft" defense in depth. To give it an offensive aspect, they used underground passageways to emerge from ground already overrun.[12] Many of the tunnel exits were so well camouflaged that they could not be found. In some areas, they were covered with live vegetation. In more barren areas, they were hidden among the rocks. Across the island, amazingly realistic dummy positions funneled the Marines into prepared kill zones.[13]

Many parts of Iwo Jima were literally covered with pillboxes.[14] Besides a direct hit from a U.S. shell, these pillboxes were at risk from two things. One was debris blocking the embrasure. This hazzard was countered by digging a ditch into which the debris could fall. The second was a grenade rolled through the embrasure. This problem was solved by digging a well into which the grenade could be kicked. While the grenade would still explode, its effects would be minimal.

Asians place a high priority on all types of deception. Defense diagrams from earlier island campaigns show what appear to be randomly placed pillboxes. On closer inspection, one can see widely dispersed, partial circles of gun emplacements.[15] In maneuver warfare terminology, these semi-independent perimeters would be called strongpoints and the undefended areas between them, firesacks. The successive lines of defense that the Marines thought they had to assault head-on in broad daylight were very probably a terrain-consistent matrix of tiny perimeters, dummy positions, and areas covered by fire. The rows of firepits that ran diagonally backwards in the ravines between automatic-weapons emplacements may have been occasionally occupied supplementary positions.[16] Where tunnels didn't exist, these strings of holes may have also functioned as avenues of egress. Kuribayashi mentioned "the establishment of 'waiting' trenches for use in close combat and raid-

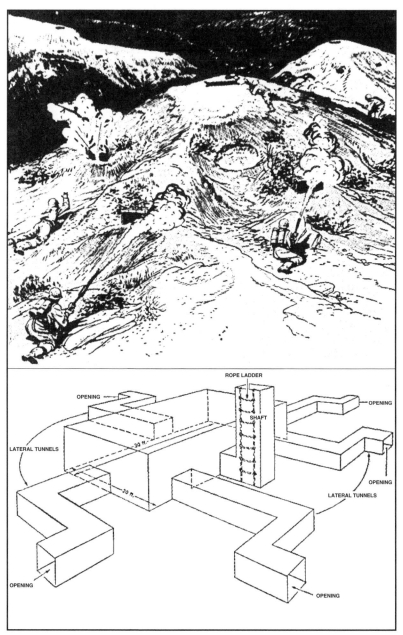

## Figure 6.4: More Than Meets the Eye

(Source: Courtesy of Osprey Publishing from U.S. War Dept. sketch in *Japanese Army of World War II*, by Philip Warner, © 1972 Osprey Publishing Ltd.)

ing" in his operations order of 1 December 1944.[17] Some of the rows of firepits may have formed inverted "V" shapes at the top of gullies. Old fighting holes on Okinawa reveal this to be the enemy's preferred method of dealing with unwanted visitors in precipitous terrain.[18] That way, the quarry has nowhere to hide after encountering fire from both sides.

The principal building block in Kuribayashi's tapestry of interlocking fire was either a well-camouflaged pillbox (possibly prefabricated) or a hollowed-out hummock of earth. Each had a machinegun and was close to a tunnel complex. In some locations, egress, resupply, or reinforcement could be accomplished entirely below ground. In other places, it had to occur through a ground-level trap door within yards of the tunnel entrance. To more easily gun down Marines from the back, the residents of some of these hardened positions may have intentionally allowed themselves to be bypassed. Others may have initially abandoned and then reoccupied their posts (the subsequent landings on Tarawa took fire from a previously captured pier,[19] and shipwreck[20]). Spider traps with metal covers may have housed forward security elements or snipers covering the ground between the strongpoints. Occupants of holes about to be overrun could have pulled shut the cover, hidden until dark, and then crawled back to their parent unit.

> The stronghold was ringed by camouflaged machinegun pits, concealed spider traps with steel covers to protect lurking snipers. Each position covered the other with lanes of crossfire.[21]

## The Underground City

One wonders how extensive these tunnel systems might have been on Iwo Jima. Because the Americans immediately sealed almost every entrance they found, there is no way of knowing for sure. Only the tunnels in the major Japanese bastions were inspected, and then only cursorily.

> Suribachi: Some 70 concrete blockhouses at its base and 50 more on its lower sides connected by tunnels to hundreds of cave entrances and pillboxes.[22]

Defense Line Number One
    The Quarry:  Coast defense guns emplaced on its rim.[23]
    Charlie Dog Ridge:  On same terrace as the Amphitheater.

Defense Line Number Two
    Minami Village
    Turkey Knob:  One cave, out of hundreds designed to be
        defended in depth, had a tunnel 800 feet long with 14
        separate exits;[24] the hill's top had a large concrete
        installation.[25]
    The Amphitheater:  Three terraces of block houses
        connected by 700 yards of tunnels;[26] on the same
        terrace as Charlie Dog Ridge.
    Hill 382:  Top of hill was hollowed out for several concrete
        artillery and antitank gun housings — each of which
        was protected by as many as ten machinegun
        emplacements — and the rest of the hill was laced with
        tunnels;[27] an elaborate tunnel system honeycombed the
        hill — nearly 1000 yards of caves for ammunition
        supply and for troops to reinforce threatened positions
        or to escape when one was overrun;[28] underground
        passageways led into defenses and when one occupant
        of a pillbox got killed, another one came up to take his
        place.[29]
    Hill Peter
    Hill Oboe
    Hill 362A:  Honeycombed with underground passages;[30]
        connected by tunnel with Nishi Ridge.[31]
    Nishi Ridge:  Some 100 camouflaged cave entrances
        interconnected by tunnels, one 1000 feet long;[32]
        connected by tunnel to Hill 362A.

Defense Line Number Three
    Hill 362C
    Hill 331
    Hill 357
    Hill 362B:  Dense network of interconnecting caves
        and pillboxes.[33]
    Kita Village:  Complex maze of pillboxes and
        interconnected caves.[34]

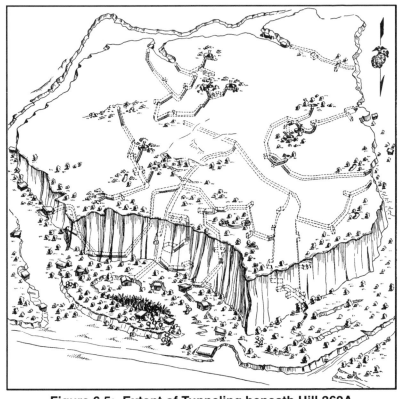

**Figure 6.5: Extent of Tunneling beneath Hill 362A**

(Source: *Iwo Jima: Amphibious Epic*, by Lt.Col. Whitman S. Bartley, from sketches by 31st Naval Construction Battalion, Hist. Branch, HQMC, p. 140)

<u>Final-Fallback Position Possibilities</u>

The Gorge and Kuribayashi's Cave: Combat engineers used 8,500 tons of explosives to detonate huge igloo-shaped concrete installation built into a knoll and circled by mutually supporting caves.[35]

Kitano Point: Entire northern tip of island was honeycombed with caves and passageways.[36]

## Amazingly Accurate Counterfire

To boggle the Western mind, an Easterner will often make a

covertly planted explosive look like a well-placed artillery shell. To spot examples of this ploy, one can only watch for the laws of probability to be violated. Anyone who has ever played "ring toss" at a county fair knows how hard it is to hit a pinpoint target.

This type of sabotage is difficult to substantiate on Iwo Jima because of the high volume of incoming fire and the extent to which the island had been preregistered by the Japanese. Still some of the nighttime and early morning "direct hits" by randomly placed shells look extremely suspicious. Could not a skilled sapper have snuck through U.S. lines, put a satchel charge and timer on something of strategic value, and then snuck back out before it was scheduled to detonate? Could not his act of sabotage have then been covered up by a few artillery or mortar shells at a prearranged time? By 1945, the Japanese had already demonstrated an uncanny ability to infiltrate American lines. How many were trying to commit suicide is pure conjecture. Detailed reports of the fighting on Iwo Jima tell of nightly infiltration attempts by three- to four-man teams.[37] While some of these attempts were being repulsed, at least four command posts (5th Corps Artillery,[38] 2/25,[39] 1/23,[40] 2/23[41]) and three ammunition/supply dumps (5th Marine Division,[42] 25th Marines,[43] 3/25[44]) were destroyed by direct hits from enemy artillery.

> For the night defense on D-plus 4 [February 23] . . . the rest of the [28th Marine] regiment manned positions around the base of Suribachi. During the hours of darkness, 122 Japanese were killed trying to infiltrate the lines. Most of these had demolitions secured to their bodies and were probably trying to reach command posts and artillery positions before destroying themselves. A few rounds of high-velocity artillery fell in the area.[45]
> — *Iwo Jima: Amphibious Epic*
> History Branch, HQMC

Soon after 2:00 A.M. [1 March], things began to change. At first, what seemed to be random enemy artillery shells started falling in a helter-skelter pattern along the western edge of Motoyama Number One [Airfield]. The area was jammed with supply and ammunition dumps, crowded with artillery positions . . . communication centers, various rear-area command posts. . . .

Mass bedlam erupted when a shell hit the main ammunition dump, touching off a chain reaction that sent explosives hundreds of feet skyward. . . .

Shortly after 4:00 A.M., when it seemed that most of
the fire was under control, near catastrophe struck when
still another explosion sent a cascade of 105-millimeter artillery shells skyward. One landed a few yards from the
communications center, which handled fire missions for
Fifth Corps howitzers. There were no casualties, but the
blast knocked out telephone lines and put out of action most
of the 105s on the island until new wires were strung to the
front.[46]

Elsewhere along the line were minor scrimmages, frequent
and heated exchanges of close-in rifle and grenade fire that
kept weary Marines awake and alert. Then, at 5:02 A.M.
[on 6 March], a large Japanese rocket made a direct hit on
the command post of the 23rd Regiment's Second Battalion.[47]

At 2300 [11:00 P.M. on 8 March] the 2nd Battalion, 23rd
Marines . . . reported large-scale infiltration attempts along
the regimental boundary in the area of the salient. . . .

. . . A few even infiltrated through the front lines to the
2/23 command post where they harassed the operations section with hand grenades thrown from ranges of 10 to 15
yards. . . .

These enemy troops were well-armed and equipped, and
many carried demolition charges. The attack was preceded
by an all-out artillery preparation.[48]

> — *Iwo Jima: Amphibious Epic*
> History Branch, HQMC

The night [of 11 March] was fairly quiet along most of the
front, but activity behind the lines of 3/27 became so heavy
that company command posts moved up to tie in with their
front line platoons for security against grenade-throwing
Japanese soldiers who came out of bypassed holes.[49]

> — *Iwo Jima: Amphibious Epic*
> History Branch, HQMC

## Antitank Mines in Just the Right Places

Since WWI, most of America's foes have had troops specially trained in killing or disabling tanks from close range.

On Iwo Jima, spider hole occupants undoubtedly practiced a well-documented Japanese technique — pulling an antitank mine on a string under the tracks of a tank.[50]

Japanese sappers darted from bypassed spider traps with sputtering demolition charges hurled against tanks.[51]

A Japanese squad rushed one tank with hand grenades that exploded in roars of green smoke. The Sherman was undamaged, but in moving to safety it plunged into a ravine and threw a tread.[52]

## Surviving the Subsequent Preparatory Fire

To withstand a bigger shell, a defender has only to dig his hole deeper. Why U.S. war planners have chosen to ignore this fact is the million-dollar question. Most military historians agree that supporting arms can pave the way for infantry only in the desert. In normal terrain, bombs and shells will have little effect on a well-dug-in adversary. The Japanese proved that once again on Iwo Jima.

The destruction and neutralization effect of supporting weapons was exploited to the fullest extent possible. However, many enemy positions withstood even large-caliber shells from ships' guns. Furthermore, the extensive networks of caves and underground passages made it possible for the Japanese to wait out a barrage secure in subterranean chambers and then come up to the surface to resist from the 200-yard zone immediately to the front of Marines, where heavy supporting weapons could not fire for fear of endangering friendly troops. For those reasons the battle was reduced to a series of close-in encounters between tank-supported Marines with flamethrowers and demolitions, and the deeply entrenched, stubbornly resisting enemy.53

> — *Iwo Jima: Amphibious Epic*
> History Branch, HQMC

The process [final push after the day of rest] began at 7:00 A.M. of March 6, D-Day plus fifteen, in thundering barrages of artillery, naval gunfire, and wave after wave of carrier planes dropping clusters of napalm bombs as their machineguns and rockets ripped into enemy lines in low-level passes across the island.  In tons of steel and explosives, the firepower approached that unleashed on D-Day in the final hour before Marines hit the beaches.

Virtually every Marine howitzer on Iwo — 132 75- and 155-millimeter guns from eleven artillery battalions — was in action.  They first fired for thirty-one minutes on the western half of the island, and then rocked the eastern sector for the next thirty-six minutes.  In sixty-seven minutes, 22,500 shells slammed down along the front, often as close as a hundred yards in front of the Marines waiting to jump off.

From less than half a mile offshore, a battleship and three cruisers unloaded 450 rounds of eight- and fourteen-inch high explosives.  Three destroyers and two landing craft moved close in along the western shoreline to plaster the cliffs and caves with mortars and rockets. . . .

. . . [When the Marine units struck,] Japanese reaction was immediate and deadly across the island.  Incredibly, the pulverizing pre-attack bombardment and air strikes seemed not to have affected the enemy at all.54

### The Deadly Embrace

Former adversaries have demonstrated another way to escape U.S. supporting arms — staying within 200 yards of their attacker.  They have done so by either letting him get extremely close before defending themselves or moving closer to him while he calls for artillery or air support.

What has been termed "the close embrace" occurred on Iwo Jima as well.  On the night of 3 March 1944, "an aggressive enemy

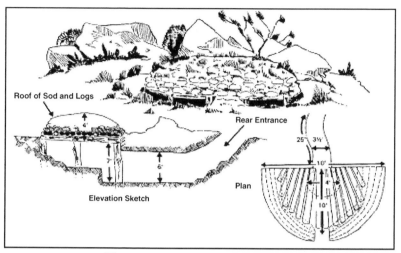

## Figure 6.6: IJA Mortar Position

(Source: *Japan's Battle of Okinawa April-June 1945*, by Thomas M. Hubler, Combat Studies Instit.,U.S.Army Cmd.& Gen.Staff College, p. 62)

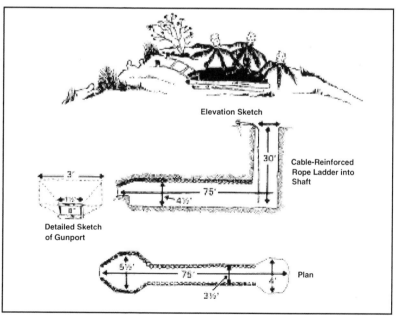

## Figure 6.7: Pillbox Cave with Vertical Entrance

(Source: *Japan's Battle of Okinawa April-June 1945*, Combat Studies Instit.,U.S.Army Cmd.& Gen.Staff College, p. 50)

crowded the 26th Marines throughout the hours of darkness,"[55] when most of the preparatory fire for the next day's attack was probably in progress.

Most of Iwo's defenders did not fanatically die in place. There is far more evidence of maneuver-war-motivated perseverance than *bushido*-related resignation. Kuribayashi's marriage of close embrace with tactical withdrawal was so ingenious as to possibly constitute a significant advance in the tactics of defensive warfare.

> The enemy remains below ground in his maze of communicating tunnels throughout our preliminary arty fires. When the fire ceases, he pushes OPs [observation posts] out of entrances not demolished by our fires. Then choosing a suitable exit he moves as many men and weapons to the surface as he can, depending on the cover and concealment of that area, often as close as 75 yards from our front. As our troops advance toward this point, he delivers all the fire at his disposal, rifle, machinegun, and mortar. When he has inflicted sufficient casualties to pin down our advance, he . . . withdraws through his underground tunnels most of his forces, possibly leaving a few machinegunners and mortars. Meanwhile our Bn CO [commanding officer] has coordinated his direct support weapons and delivers a concentration of rockets, mortars, and artillery. Our tanks then push in, supported by infantry. When the hot spot is overrun, we find a handful of dead Japs and few if any enemy weapons. While this is happening, the enemy has repeated the process and another sector of our advance is engaged in a vicious firefight, and the cycle continues. Supporting indications to these deductions are:
>
> (1) When the hot spot is overrun we find far too few dead enemy to have delivered the fire encountered in overrunning the position;
> (2) We find few if any enemy weapons in the positions overrun but plenty of empty shell cases;
> (3) We find tunnel entrances, some caved in, all appearing deep and well prepared, some with electric light wires;
> (4) During the cycle, close air and OP observation detects *no* enemy surface movement.[56]
>
> — 4th Mar.Div. Intelligence Report of 6 March 1945

## Covered Avenues of Reinforcement and Egress

Iwo Jima's labyrinth of passageways was almost certainly more extensive than discovered at the time, many of its entrances were never found. It is known that Hill 362A and Nishi Ridge were connected by tunnel.[57] Who's to say that a tunnel did not link the strongholds in each of the two northernmost defense lines, or that a north-south conduit did not connect those defense lines with their counterpart south of Airfield Number One? Could not the Japanese have sufficiently disguised below-ground entrances and blown up threatened connector tunnels to keep the existence of a central thoroughfare secret? It is known that they blew up captured sections of Nishi Ridge and other bastions.[58] Wouldn't they have needed such a conduit to share limited resources and forage behind U.S. lines? After all, the Japanese only had so much water and ammunition.

The Nipponese were also skilled at above-ground infiltration. While Suribachi was intermittently illuminated every night, many of its survivors may have snuck through Marine lines to join the fight in the north. A document found in Suribachi's crater proved that some had tried.[59] This might explain why the volcano's slope had been conquered by three Marine scouts after its base had been unapproachable for days,[60] and why bypassed defenders were strangely inactive after the flag raising.[61]

But Iwo was not just a well-fortified island; it was an island on which the Japanese could change location without ever returning to the surface of the earth. The implications are staggering: (1) the island may have been defended by far fewer troops than is commonly acknowledged, and (2) this may have been an important refinement to the Germans' flexible defense.

Marine spotters in Maytag Messerschmitts watched the attack come to life in brilliant, chilly sunlight. Theirs was a familiar spectacle on Iwo, but totally foreign to anything else in military history. Below them was a battlefield where one army fought above ground and the other fought almost totally beneath it; where thousands of troops moved in the same area at the same time, Marines maneuvering on the surface in the attack, the Japanese moving in tunnels from one underground strongpoint to another whenever the tide of combat called for more firepower or reinforcements.

Everything beyond the line was completely barren and devoid of any sign of enemy troops — a rugged, uninhabited landscape as mysterious as a solar planet.[62]

Some of the machinegun and sniper attacks from the rear may have come from reoccupied positions. In heavy combat, leaders will often ignore single shots from behind thinking them random or from disgruntled subordinates. To partially contain this threat, large forces had to follow the lead regiments into combat.[63] Even then, the lead units received heavy fire from bypassed positions.[64]

"Every hillside, every ravine, had its camouflaged cave or pillbox; some were so carefully hidden that men stepped on them before they were aware of them," his copy said. "One cave in the Fourth Division area, northeast of the first airfield, had a tunnel eight hundred yards long with fourteen entrances. Each entrance was covered by a series of pillboxes containing machineguns. If the inmates of one pillbox were killed, the Japs could easily send out replacements from another entrance. Japs would pop out of holes in the ground far behind our own lines."[65]
— Robert Sherrod
*Time-Life* correspondent

The island, merely eight-square miles, was riddled with a sophisticated system of enemy tunnels. Moving from one objective to another, Marines were easy targets for the Japanese who were able [to] duck into this "basement of catacombs" and maneuver behind the American troops, shooting them in the backs.[66]
— *Camp Lejeune Globe,* 8 October 1999

Not only had eleven battalion commanders been killed or wounded since D-Day, but the loss of junior officers and senior noncoms at the company, platoon, and squad levels had been devastating. . . .
More than two thousand replacements — mostly second lieutenants new to combat and men fresh from the States — had been thrown into the struggle.[67]

In a subterranean cross-section of the island from a Japanese television documentary, most of the island is spanned by a string of interconnected underground rooms. The thoroughfare extends from just east of Airfield Number One to the northeastern shore. The scene in the next frame resembles a glassed-in ant hill. It shows what the average bastion looked like below ground. There are some interesting details: (1) surface hummocks hollowed out as gun emplacements, (2) an interconnecting tunnel network with other cave entrances, and (3) several levels of subterranean rooms. Some of the lower spaces appear to have false floors obscuring large ammunition and supply chambers below.[68] This might explain why it was so easy to cross the low ground south of Airfield Number One and difficult to move north from there. It is known that Kuribayashi had planned to connect his emplacements with 38,000 meters of underground passageways.[69] By the start of 1945, 26,000 meters of tunnel had been completed.[70] This was more than enough to link the key bastions. The combined length of the three defense lines was only 14,000 meters. The distance between Airfield Number One and the island's northern tip was 5000 meters.

Kuribayashi demanded the assistance of the finest mining engineers and fortifications specialists in the Empire. Here again the island favored the defender. Iwo's volcanic sand

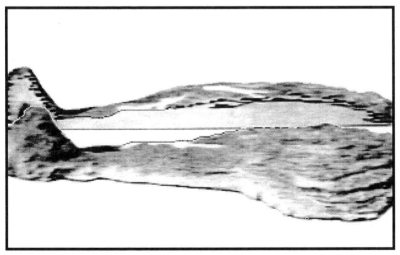

**Figure 6.8: Cross-Section Shows Tiers Connected by Tunnel**
(Source: *A Tribute to WWII Combat Cameramen of Japan,* Nippon TV videocassette)

mixed readily with cement to produce superior concrete for installations; the soft rock lent itself to rapid digging. Half the garrison lay aside their weapons to labor with pick and spade. When American heavy bombers from the Seventh Air Force commenced a daily pounding of the island in early December 1944, Kuribayashi simply moved everything — weapons, command posts, barracks, and stations — underground. These engineering accomplishments were remarkable. Masked gun positions provided interlocking fields of fire, miles of tunnels linked key defensive positions, every cave featured multiple outlets and ventilation tubes. One installation inside Mount Suribachi ran seven stories deep. The Americans would rarely see a live Japanese on Iwo Jima until the bitter end.[71]

*— Closing In: Marines in the Seizure of Iwo Jima*
History and Museums Division, HQMC

**Figure 6.9: Rooms below the Hummocks**
(Source: *A Tribute to WWII Combat Cameramen of Japan*, Nippon TV videocassette)

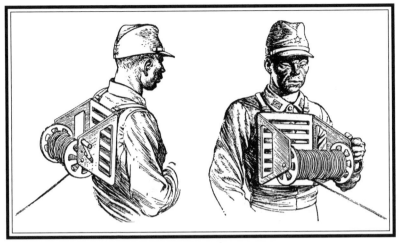

**Figure 6.10: Underground Communications**
(Source: *Handbook on Japanese Military Forces*, TM-E 30-480 (1944), U.S. War Dept., p. 321)

How much of the connector tunnel had been completed by D-Day may never be known. There is evidence that Major General Senda somehow moved from his brigade headquarters just east of Hill 382 to the 4th Marine Division's final pocket — some 1200 meters to the southeast.[72] This would indicate the eastern bastions of the second defense line might have been connected by tunnel as well.

It is also known that every piece of key terrain on the island had its own elaborate tunnel system,[73] and that the Japanese could dig at a rate of 1/3 meter per hour.[74] During the 35-day battle, a single team could have tunneled 300 meters. There is no telling what hundreds of teams working toward each other from different locations might have accomplished.

## The Unexpected Counterattack

The proponent of maneuver warfare will often try to recapture lost ground, but his counterattack may not be immediate. After all, to be too predictable or noisy is to fail.

The counterattacks on Iwo Jima were not of the *banzai* variety that the Marines had come to expect from the Japanese. Many of them were by fire alone from distant locations. Against the heavily concentrated Marines, they were deadly. Others were well-planned and quietly executed raids. The last of these raids gives further credence to the hypothesis that a secret tunnel spanned four fifths of the island. On 26 March, two weeks after Kuribayashi's force had been supposedly destroyed at the north end of the island,[75] something amazing happened. Three hundred Japanese attacked Airfield Number One.[76] POWs (prisoners of war) later claimed that Kuribayashi had personally led the assault.[77]

## The Bottom Line

When an attacker measures his success by body count instead of strategic contribution (as has been the U.S. tradition), his self-esteem largely depends on how many dead enemy he can find. Not many Japanese corpses or POWs were retrieved on Iwo Jima. It is generally thought that most of the defenders were trapped below ground when their cave entrances were sealed. Still somewhat disturbing is the Japanese diagram showing multiple tiers of subterranean rooms at the northeastern end of the island (look again at Figure 6.8). Could General Kuribayashi have planned one final deception? Did those who had manned the southern defense lines pull back to "Kuribayashi's Cave" or somewhere else? Could hundreds have hidden long enough to be incrementally withdrawn by submarine? Near the end of the war, the Japanese Army did have its own cargo submarines.[78] And Kangoku was only one of several sets of rocks that lay less than a mile offshore.

According to U.S. records, the number of estimated defenders on Iwo Jima rose by half over the course of the battle.[79] Some of those killed or captured must have come from units not previously known to be on the island. Still one wonders just how many Japanese had really been there. Two months earlier at the Battle of the Bulge, the retreating enemy had been pummeled with supporting arms until the casualty ratio shifted in favor of the Allies.[80] Luckily, for the Marines who paid such a terrible price for each square yard of real estate on Iwo Jima, the island as a whole had tremendous strategic value. It would greatly facilitate the bombing of Japan and retrieval of damaged aircraft. Some 25,000 U.S. airmen may owe their lives to their determined brethren in arms.

# 7     The Human-Wave Assault Deception

● *How risky is the Eastern version of the mass assault?*

● *Which forces win after the ordinary forces engage?*

(Source: Courtesy of Cassell PLC, from *World Army Uniforms since 1939*, © 1975, 1980, 1981, 1983 by Blandford Press Ltd.; Corel Gallery, *Clipart* — Plants, 35A031)

## The Eastern Mass Assault Rationale

On occasion since 1942, America's Asian adversaries have resorted to what looks like a WWI vintage, "human-wave" assault. In fact, it is to this maneuver that they owe much of their reputed disregard for human life. In war, it is the overall casualty count that matters most. Many soldiers must sometimes be sent into harm's way to preclude the loss of many more later. The Eastern group rush is not very different from a Western-style amphibious landing. Might not both be appropriate under the proper circumstances? Routinely overprotecting one's infantrymen can actually do more to raise the final casualty count, because it undermines momentum. Without maneuver force momentum, wars take too long.

Being highly deceptive, the Asian may have misled his traditional adversary by periodically mimicking the successive waves of Allied infantrymen who stormed enemy strongpoints throughout WWI. If the Asian uses "ordinary" forces to engage his opponent and "extraordinary" forces to beat him,[1] might not the human wave qualify as an example of the former? While obsessed with enemy hordes to the front, might not a Western defender pay too little attention to his rear? By running the occasional mass assault, the Easterner may further cause his well-supplied and attrition-minded adversary to underestimate what lone infiltrators and stormtrooper squads could do to his command-and-control, logistics, and fire-support infrastructures.

After all, extraordinary-force contributions can be made to look like aborted probing attacks, lucky mortar hits, or routine accidents. The Western commander may overlook what is less threatening than the prospect of being overrun. He has been taught to gauge his success by the casualty ratio for each encounter. He concentrates more on safely escaping those encounters than on furthering the overall war effort. Unfortunately, his final casualty count depends more on the war's duration. Failing to comprehend what individual soldiers and small units can accomplish strategically, Western commanders may be tempted to undertrain and overcontrol them. This does little to enhance their chances of surviving one-on-one encounters with their Asian counterparts. In a cruel paradox of war, the Western commanders will lose many of the people they have been trying too hard to protect.

## The Oriental Version of the "Human-Wave" Assault

While still inherently dangerous, the Asian mass assault may be quite different from its Allied forerunner. It may have been subtly changed to minimize the risk. Perhaps a way to lessen the machinegun threat has been added. Would not a barrier of interlocking automatic-weapons fire only hurt the people who tried to walk or run through it? The Limies, Frenchies, and Doughboys of WWI were expected to remain upright throughout each assault. Possibly, the Japanese, Chinese, and North Vietnamese were not.

They may have been allowed to crawl part of the way or to dive over or under the narrow streams of tracer bullets. They regularly practice acrobatic moves and have been recently photographed "flying" five feet above the ground.[2] Moreover, the Allied assault troops of WWI had to contend with continual illumination and barbed wire. Eastern armies have only attacked *en masse* when both obstructions were minimal.

Most importantly, the Eastern commander may have enjoyed more opportunity than his Western counterpart to minimize the cost of such an attack. Instead of an established scheme of maneuver, he may have had interchangeable components of fire, movement, and deception. Perhaps, his subordinates had an edge as well. Instead of being required to stick to a preconceived game plan, they may have been able to shift between prerehearsed techniques in mid-assault. A closer look at the Asian-style human-wave attack could reveal a tactic much less wasteful than commonly believed. To uncover its subtleties, one must establish a working hypothesis.

*Planning*

The Eastern commander requires his infantry squads to heavily rehearse several interchangeable assault techniques before leaving their base camp.

Then, he has several ways in which to reconnoiter an objective: (1) infiltration days before, (2) close-range observation 24 hours before,[3] (3) probing right before,[4] and (4) "reconnaissance pull" (exploiting local successes) during the attack.[5] He does so to locate gaps, unit boundaries, and machinegun emplacements.

*The Final Stages of the Approach March*

On the night of the attack, several successive rows of Asian assault troops silently approach Western lines.[6] In alternating rows are burpgunners and grenadiers.[7] Each row consists, not of a separate unit, but of squads from different companies walking abreast. Perfectly camouflaged to look like the surrounding countryside, the Oriental troops remain motionless under illumination.[8] If the first wave encounters a listening post, it attempts to sneak a few people into its objective on the heels of withdrawing sentries.[9] If it does not encounter a listening post, it attempts to crawl undetected to within grenade range of opposing lines.[10]

### The Initial Assault

If the lead element can closely approach their quarry, no fewer than four waves participate in the initial attack. In that way, whole platoons can quickly enter any penetration. The first wave assaults silently — without any shouting, shooting, or explosives[11] — or throws concussion grenades in support of succeeding waves.[12] During that assault, the Oriental infantryman knows how to safeguard the element of surprise and his life in the process. He knows that the next best thing to total silence is what sounds to defenders like impacting mortar rounds. He bayonets rather than shoots his immediate adversary,[13] and then uses concussion rather than fragmentation grenades to expand his holdings.[14] Once the opposition machineguns open up, he has less reason to withhold his small-arms fire.

If, for whatever reason, the lead element cannot closely approach, the first wave comes in throwing concussion grenades behind a rolling mortar barrage,[15] or shooting burpguns under overhead machinegun fire.[16]

### The Subsequent Grinding Action

After the initial mass assault, the follow-up attacks are conducted piecemeal — at different times and from various directions. Although squad sized, some appear larger. The leader of the next squad in each lane decides when to attack. Sometimes he sees strength in numbers; other times, a distant squad's assault as a helpful diversion of enemy attention.

Succeeding squads from the same company have options. They can do more than just take turns — throwing grenades and firing small arms[17] — until their assigned sector is reduced.[18] Their members are encouraged to show initiative as well. Those who get in assist others by grenading,[19] or shooting,[20] frontline defenders from the rear.

In fact, each squad has a veritable "portfolio" of prerehearsed "plays." Its members can fire burpguns without advancing. By shouting and shooting from covered positions, they are able to gather the latest intelligence on defender strength. Or they can throw grenades while crawling, instead of running, forward.[21]

Because the sister squads have practiced ways to cooperate, this relentless pounding on one small sector can be extremely powerful. Like the Japanese squads on Guadalcanal,[22] the first can throw concussion grenades, while the second rushes in shooting.[23] Or, from behind rocks or stumps, the first can fire over the heads of a crawling second. In essence, any pair of squads can perform the first two parts of a maneuver warfare attack — penetration and suppression. The rest of their company is well positioned to perform the third — exploitation.[24]

To make matters worse, this grinding action has eyes. Imbedded in the Oriental technique is a form of "recon pull." Each succeeding squad has the opportunity to learn from its predecessor — to, at the last minute, pick the most appropriate of several prerehearsed techniques. Its company commander can track and report its relative progress. In turn, the battalion commander can shift his focus of main effort in mid-attack — reassigning portions of companies to lanes offering the least resistance.[25] In essence, the random, squad-sized attacks uncover weak spots that can be exploited.[26]

### The Secret Initiative

All the while, the Asian's "extraordinary" forces are sneaking into the objective from the rear. Initially, they go after the defenders' command bunker, mortar pit, and ammunition dump with grenades.[27] Then they help their "ordinary" comrades to break through. They are the source of some of the grenades and small-arms fire with which the frontline defenders are inundated. Simultaneous fire from front and back is difficult to differentiate.

### The Withdrawal without Consolidation

At a prearranged signal, human wave participants melt back into the night — often to regroup for the next round of attacks.[28]

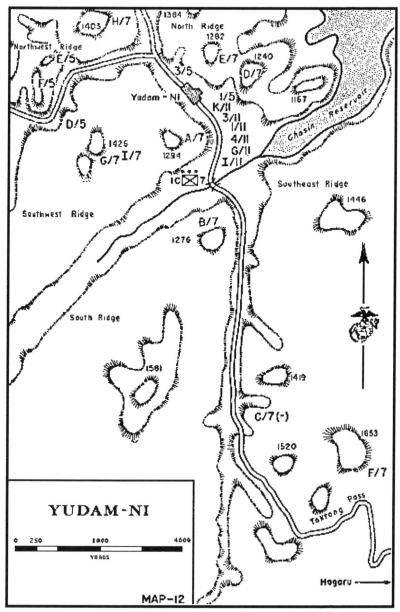

**Map 7.1: Yudam-Ni Valley South to Toktong Pass**

(Source: *The Chosin Reservoir Campaign,* by L. Montross and N.A. Canzona, vol. III, U.S. Marine Operations in Korea: 1950—1953, Hist. & Museums Div., HQMC, p. 153)

By more closely examining this "progressive" form of penetration, a Westerner might discover how to lower the cost of frontal attack. Massed Chinese assaults occurred against several U.S. company-sized outposts on the same frigid night in November 1950.

## Setting the Stage

Through the careful camouflaging of personnel and setting of forest fires, no fewer than ten Chinese divisions had managed to infiltrate into North Korea without being noticed by U.S. reconnaissance aircraft. As two regiments from the 1st Marine Division advanced westward through the village of Yudam-ni at the southern end of the Chosin Reservoir on 27 November 1950, they encountered long-range fire from the high ground to their front. They moved a little further west and then outposted the hills along the valley's edge.

> When dusk came at around six o'clock, the mass of the [M]arines had moved off the valley floor and into the hills: ten understrength rifle companies on the high ground; two battalions of the 5th Regiment in the valley near the village [of Yudam-ni]; and two rifle companies, Charlie and Fox of the 7th Regiment in isolated positions along the [south-bound] road to Hagaru.[29]
> — *U.S. Marine Operations in Korea*
> History Branch, HQMC

Unbeknownst to the Marines, the 59th Chinese Division was about to move against the 14-mile stretch of road to Hagaru,[30] the 79th the southern end of North Ridge, and the 89th the southern end of Northwest Ridge.[31]

## The Battle of Northwest Ridge

The first enemy thrust came down Northwest Ridge three hours before midnight on 27 November. While targeting Fox and Easy Companies of the 5th Marines in the western foothills, it began with a simulated mortar attack against another unit in the valley. Then came the standard probing of lines and concentrated assault on a weak spot. To this assault, the Chinese added overhead machinegun fire and a rolling mortar barrage.

The first serious flare-up, a diversion, occurred when a group of Chinese mounted a grenade assault against a roadblock manned by a platoon of Dog/5. . . .

When attention was drawn to the roadblock fight, masses of PLA [People's Liberation Army] infantry maneuvered to within yards of the main line held by Fox and Easy companies. . . . [L]ight probes were launched by very small Chinese groups along the Fox/5 line. The Chinese recoiled wherever they met resistance, but by drawing fire they exposed the locations of . . . Marine automatic weapons.

While attention was drawn to the light probes [elsewhere], infiltrators intent upon breaching the Marine line crawled to within a few feet of the junction of Fox and Easy Companies. . . .

This was a skillful series of ruses and deployments. . . . Satisfied in time that he had found all that could be revealed, the PLA commander ordered his assault. . . .

[G]renadiers hurled bunches of concussion stick grenades while machineguns on the heights . . . probed the night sky with . . . green tracer. . . . [A] sustained mortar barrage caught many Marines in the open, on the move.[32]

At 2125 [9:25 P.M.] the mortar eruptions began to walk toward the Marine rear. Whistles screeched, enemy machineguns fell silent, and the first Chinese assault waves hurled themselves against the juncture of Companies E and F. The enemy attacked on an extremely narrow front [initially] in order to maintain control. His troops advanced in column within grenade range, then deployed abruptly into skirmish lines that flailed the Marine positions ceaselessly. . . .

Ultimately, the Reds broke through . . . where the two units were joined. They poured troops into the gap, and as they attempted to roll back the newly exposed flanks, they

overran part of Fox Company.[33]
— *U.S. Marine Operations in Korea*
History Branch, HQMC

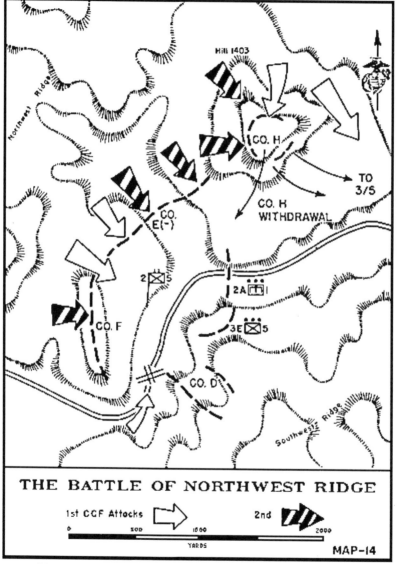

**Map 7.2:  The Chinese Attack down Northwest Ridge**

(Source:  *The Chosin Reservoir Campaign*, by L. Montross and N.A. Canzona, vol. III, U.S. Marine Operations in Korea: 1950—1953, Hist. & Museums Div., HQMC, p. 162)

> Massed PLA infantry, led by ranks of submachine-gunners, tore into the Marine lines, the forwardmost three hundred of them driving a wedge at the point where Fox and Easy Companies were joined. . . . [T]he Chinese went right after the machineguns supporting the right Fox Company platoon. . . .
> . . . First there would be an attack by grenadiers, then a lull, then an attack by submachinegunners, then a lull, then another grenade assault.[34]

The assaults up front came from different directions and ominously whenever opportunities arose.

> There was a lull in the fighting, but there was no telling how long it might last, nor where the Chinese would strike next . . .
> [T]he nearest heavy machinegun quit firing, and the Chinese mounted another attack.[35]

While this was occurring up front, infiltrators entered the U.S. position from the rear. They were trying to destroy U.S. mortars to curtail illumination, and U.S. leaders to diminish control.[36]

> A Chinese infiltrator shot at Sergeant Johnson as he stood in the momentary flash caused as a round left the [81mm mortar] tube he was steadying. The big noncom was thrown onto his back, the shock of a bullet through the groin killing him within minutes.[37]

> [A bazooka] rocket sergeant [had earlier been] . . . shot by an infiltrator hiding on the rear slope of the ridge.[38]

The Chinese were to come back against the center of the Fox Company line several hours later, inflict many casualties, and overrun two machinegun positions.[39]

## Details of the Technique Emerge

The Chinese had identified and attacked a narrow, lightly defended unit boundary and then sent reinforcements through the breach — a classic example of small-unit "maneuver warfare." As the Japanese before them, they may have tried to funnel too many reinforcements through each hole.

Many Asians had participated in the assault that night, but not all had experienced heavy counterfire. They did not mechanically advance as the waves of Allied soldiers had in WWI. Their approach had been made in parallel columns of companies.

> Unseen and unheard [after sunset on 27 November], the endless *columns* [italics added] of quilted green [actually white] wound through valleys and over mountain trails leading toward the southern tips of Northwest and North Ridges. These were the assault battalions of the 79th and 89th CCF [Chinese Communist Forces] Divisions.[40]
> — *U.S. Marine Operations in Korea*
> History Branch, HQMC

> [CCF] regiments, attacking in columns of battalions deployed in columns of companies.[41]

In the Chinese approach march, skilled infiltrators — possibly regimental/battalion scouts — led the way. They may have had as their subsequent mission to enter the objective from the rear.

> The first *files* of Chinese *skirmishers* [italics added] crept down from the heights opposite the slumbering Marine lines along the northwest, north, and northeast arcs of the Yudam-ni perimeter.[42]

> It [3rd Battalion, 236th Regiment, CCF] . . . sent out the usual screen of infiltrators.[43]
> — *U.S. Marine Operations in Korea*
> History Branch, HQMC

At the front of each battalion had been a squad specially trained in probing lines.[44] As with the infiltration attempt, their activities had probably been coordinated from above. In addition to locating gaps and automatic weapons, their job may have been to divert U.S. attention from the sectors to be infiltrated or assaulted.

The first activity . . . was noted . . . when several PLA squads approached. . . . Light skirmishing ensued for about thirty minutes, in which time the probers were driven off at a cost of three Marines wounded.[45]

As the first few squads in each column had neared the U.S. position, they had deployed into skirmish lines and come roughly abreast with squads from other companies. Collectively, the various squads in each row looked like a cohesive unit on line — the Allied method from WWI.[46] While momentarily exposed, those Asian infantrymen could have done worse. They could have stayed in column near an opposition machinegun. Those in the first row could now defend themselves without shooting each other in the back. If the lead squad could not get in, it had help close at hand. The next squad was in a perfect position to continue the assault or provide covering fire. If the two squads had prerehearsed ways to cooperate, their combined efforts might have been quite effective.

In each lane, the subsequent attacks were probably randomly launched whenever the defenders' attention had been diverted, illumination had died down, or machinegun had paused (presumably to be reloaded or unjammed). Coming directly at a Marine machinegunner in total darkness is not as dumb as it sounds. The gunner has normally been assigned a narrow direction of fire to his left or right. His job is to protect an adjacent sector of the lines, not his own.[47] The Asian soldiers had only to rush the gunner whenever he stopped firing. And, to keep from being too predictable, their squad leaders had several prerehearsed assault techniques from which to choose — each with its own deception with which to rebuild the element of surprise.

Finally, the rest of the squads in each company column may have been exposed only to infrequent mortar and artillery explosions.[48] To escape machinegun fire in this type of terrain, one has only to hug the ground. The North Korean countryside is rugged — covered with rocks, stumps, and depressions behind which to take cover. Machinegunners often shoot high at night. The Japanese had reestablished that in Manchuria in 1939.

Again the Soviets opened a violent storm of automatic-weapons and heavy-machinegun fire and flung grenades at flickering shadows. Both Japanese platoons again withheld any counterfire and continued to advance silently, guiding on their platoon leaders' backs. When they [the

Japanese] had closed to within a few meters of the Soviet front line . . . , they leaped into the enemy trenches slashing and bayoneting the Soviet defenders there.[49]
— Leavenworth Papers No. 2, U.S. Army

## Back to the Fight on Northwest Ridge

While Fox Company of the 5th Marines was being hit at the southwestern end of Northwest Ridge, How Company of the 7th Marines was getting the same treatment at the southeastern end. As one side of the Marines' position on Hill 1403 was being probed by squad-sized Chinese patrols, the other side was being noiselessly stalked by hundreds of enemy infantrymen.

Captain Cooke checked in [over the land line] moments before 2200 [10:00 P.M.], informing Battalion that How Company had been subjected to light probes by small PLA patrols during the whole of the preceding thirty minutes.
No sooner said than the Chinese mounted a massed assault against the two-platoon line atop the hill.[50]

Most of the white-clad stalkers may have rushed in without throwing grenades or shooting. One contingent started yelling at the last moment to startle and thus better locate their victims.

A full company of Chinese glanced off the edge of the fight atop Hill 1403 and streamed down the slope in the direction of the pup tents. . . . [The] Chinese moved with great stealth upon the former campsite.
The attackers rushed forward on command, screaming to awaken and confuse reposing Marines, thrusting bayonets into what they thought would be bodies in the empty sleeping bags.[51]

Those who penetrated U.S. lines (from whatever direction) may have had as their subsequent mission to destroy U.S. machineguns, command bunkers, mortar pits, or ammunition. To do so unnoticed with small-arms fire would have taken a well-thought-out deception.

As the left platoon braced to take the brunt of the assault,

**95**

a lone Chinese burpgunner infiltrated to the Marine rear and squeezed off telling bursts as he scrambled from position to position to escape detection.[52]

After a two-hour respite, the Chinese came back after How Company. They started with the same diversion they had used against F/2/5 and then hit with concussion grenades. Assault wave segments alternated between grenades and shooting until the Americans were finally ordered to withdraw from their hill.

While the roadblock [in the valley] was being hit, the Chinese came back against How Company, beginning with the usual shower of concussion grenades. . . .

The Chinese were all over the dwindling perimeter, too close and too intermingled to be dealt with in any way but hand-to-hand. *The [Asian] attackers were more intent upon breaking through How / 7 than in destroying it*, and [several] surprised Marines emerged from losing scuffles as their adversaries scrambled over the crest and descended toward the valley below [italics added].[53]

## The Battle for North Ridge

At about the same time elements of the 89th CCF Division were conducting assaults down Northwest Ridge, parts of the 79th Division were descending on the two isolated companies of 2/7 at the southern end of North Ridge.[54] Though lost, the Chinese had the versatility to continue to operate.

The 79th PLA Division was deployed to seize three of the four hills guarding the western side of the Yudam-ni base. . . . [E]ach of the division's three regiments was assigned an objective that did not appear to Chinese observers to have been occupied by American Marines: the rightmost Chinese regiment was to seize Hill 1384, *behind* which Taplett's battalion [3/5] had come to rest in the late afternoon of 27 November. The center Chinese regiment was to take Hill 1240, *behind* whose summit Milt Hull's Dog/7 had been camped since November 26. The leftmost Chinese regiment was to take Hill 1167, which was not occupied at all by Marines. Only Hill 1282, between Hills 1384 and 1240, was to be spared. . . . It seems that the commander of the 79th PLA Division had decided to move into the valley of Yudam-ni against the least possible opposition, by way of the "undefended" heights.

The forbidding terrain knocked the Chinese plan askew. The regiment bound for Hill 1384 found its way, but the two southern regiments . . . veered northward. Thus, unoccupied Hill 1167 was not assaulted; the regiment bound for it moved on Hill 1240, and the regiment bound for Hill 1240 blundered toward Hill 1282. While this placed both Marine companies in danger, the Chinese advantage of freedom of movement was negated by the fact that the [assault] troops would be delivering their attacks across totally unfamiliar terrain, at night, against unanticipated opposition.[55]

Unlike the other companies, Easy on Hill 1282 knew in advance that the Chinese were interested. The night before, an alert sentry had grenaded a Chinese officer mapping their position.[56] Right at 10:00 P.M. on 27 November, began two distinct rounds of probing. The second occurred in the sector about to be subjected to human-wave assault.[57] Squads containing both grenadiers and burpgunners would hit the lines and then recoil. Then about midnight, the Chinese were heard crunching through the snow to the cadence of a single chanting voice.[58] As illumination went up, the Marines saw what appeared to be four endless rows (15 yards apart) of enemy soldiers.[59] When the defenders started shooting, the Chinese rushed forward throwing concussion grenades and blowing whistles and bugles.[60] Then began a period of consecutive assaults at different locations along the line.[61] The Chinese appeared to be employing random squad rushes from different directions.

Resorting to grinding tactics, the Chinese repeatedly assaulted Company E's position from midnight until 0200 [2:00 A.M.] . . . [with] charging squads of infantry.[62]
— *U.S. Marine Operations in Korea*
History Branch, HQMC

By 2:30 A.M., Easy had been only partially overrun, so the

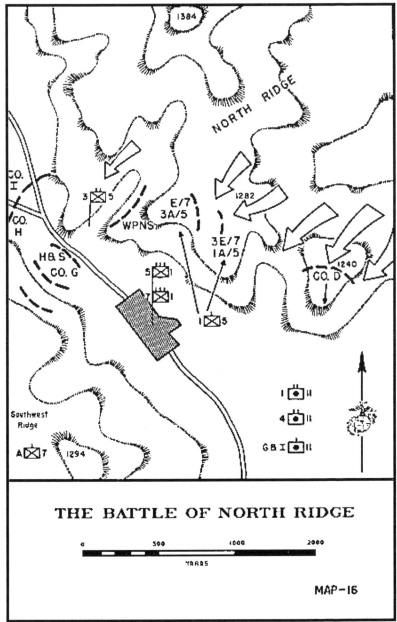

**Map 7.3: The Chinese Attack down North Ridge**

(Source: *The Chosin Reservoir Campaign*, by L. Montross and N.A. Canzona, vol. III, U.S. Marine Operations in Korea: 1950—1953, Hist. & Museums Div., HQMC, p. 173)

Chinese withdrew. Recall must have been an integral part of their technique. Whenever the bugle blared, the Chinese would cease fighting and pull back.[63] Western forces have no such signal — once committed, they are expected to finish every attack.

By this time, two platoons from A/1/5 had been dispatched from the valley floor to help out on Hill 1282. Before they could arrive, one flank of Easy Company and then the other was being attacked by the Chinese.[64] Again random squad-sized assaults were in evidence.

> In squads of eight to ten, the Chinese struck again and again at the perimeter on the summit, and the two depleted platoons of Easy Company dwindled to a mere handful of tired, desperate Marines.[65]
> — U.S. Marine Operations in Korea
> History Branch, HQMC

Easy's command post was eventually overrun, but then its reserve 3rd platoon took it back. When the valley relief force finally arrived, part was fed into the lines as replacements and part assumed flank security.[66] But the Chinese hit again, and this time they were too strong for what remained of the combined force.[67]

> The Reds finally drove a wedge between the Marine defenders on the summit and the platoons . . . on the [mountain] spur. . . .
> By 0500 [5:00 A.M.], CCF infantrymen . . . occupied the summit of Hill 1282, still believing it to be Hill 1240. . . . The remnants of the [Marine] platoons . . . had been driven to the reverse slope in the west. . . . Up to this point, Chinese casualties on Hill 1282 probably numbered about 250, with Marine losses approximately 150. Easy Company had been reduced to the effective strength of a rifle platoon (split in two), and the pair of A/5 platoons paid with upwards of 40 killed and wounded during the brief time on the battle line; only six effectives remained in [one platoon].[68]
> — U.S. Marine Operations in Korea
> History Branch, HQMC

Meanwhile on Hill 1240, Dog/7 had been getting about the same treatment as Easy/7. Shortly after 10:00 P.M. on the night of 27

November, it experienced an extended period of light probing. At 1:05 A.M., the Chinese launched a full-fledged assault.[69] Succeeding waves alternately hit different parts of the line.[70] Before long, infiltrators were tossing grenades at Marine mortarmen.

> [At] 2345 [11:45 P.M.], Company D of 2/7 reported enemy infiltration on Hill 1240.[71]
> — *U.S. Marine Operations in Korea*
> History Branch, HQMC

First Lieutenant Bill Schreier, the company mortar officer, was directing his crews amidst exploding hand grenades and mortar rounds when he glanced up to see a half-dozen PLA infantrymen coming right at him. He snapped his carbine up and fired, stopping the attackers momentarily, until the simultaneous explosions of numerous grenades forced him to duck. Schreier next saw about twenty Chinese heading his way. His fire had little or no effect, so he trundled uphill to the company command post where he found the wounded company commander.[72]

By 3:00 A.M., the two forward platoons had been overrun and the rear platoon ordered to reorganize at the foot of Hill 1240.[73] The Marines counterattacked but then succumbed to another Chinese onslaught. At dawn only isolated pockets of wounded Marines remained on the rear slope of Hill 1240.[74]

> Dog Company held, broke, counterattacked, broke again, and finally was smashed by overwhelming numbers of Chinese. When dawn came, Hull had been wounded several times, and he had only sixteen men left in fighting condition. The Chinese stood in front of him, on higher ground; on both his flanks; and on the slopes to his rear.[75]

### Then There Was Toktong Pass

Finally came the turn of the company guarding Toktong pass — the highest point on the road between the 1st Marine Division maneuver force in the Yudam-ni valley and their advance supply base at Hagaru-ri. Fox Company, 2/7, reinforced by 81mm-mortar and heavy-machinegun attachments, guarded a hill just north of the road. At about 2:30 A.M. on 28 November, rows of white-clad Chinese infantrymen came rushing across the moonlit saddle connecting Fox Hill to the much higher Toktong-San. The first wave may have silently stalked their quarry, advanced under cover of overhead machinegun fire (which was subsequently shut off by bugle[76]), and then assaulted firing their submachineguns.[77] Their assault carried the topographical crest of the hill.[78]

Five minutes after the ordinary forces had attacked from the front, a small column of Chinese advanced on the company command post from the rear.[79] Upon being seen, the phantom force went after the Marine 60mm mortars with grenades.[80]

Then the assault troops shifted their efforts to what had been a gap between 2nd and 3rd Platoons and replaced their small-arms fire with concussion grenades.[81] The use of grenades by both front- and back-door forces may have been no coincidence. At night, a grenade's trajectory is difficult to ascertain. Then the Chinese bugles sounded recall, and the assault troops melted back into the darkness. They were content to conduct small probes for the rest of the night.[82]

## The Final Outcome

The Chinese mass assaults of 27 November 1950 may have accomplished much of what was intended. They had been executed in 20-below temperatures, and by 1:15 A.M., full moonlight.[83] Though totally disoriented, the enemy regiments assaulting Hills 1240 and 1282 had still managed to take both.[84] Other regiments had not only displaced How/7 from Hill 1403 on the southeastern end of Northwest Ridge, but also Charlie/7 from Hill 1419 on the road to Hagaru,[85] How/5 from Hill 403 at the head of the Yudam-ni valley,[86] and elements of Item/5 from Hill 1384 at the southwestern end of North Ridge.[87] How many more Marine companies were temporarily split so that Chinese forces could pass through may never be known. Most of the attackers were not trying to kill defenders nor hold high ground, only to break through to the valley floor. Not all of the reconstituted lines had held long enough to thwart these intentions. While not annihilated on the night of

27 November, U.S. regiments involved were still forced to alter their strategic goals. Still, the Marines did well based on what they had been told about their foe.

The Chosin Leathernecks had limited barbed wire and indirect-fire support, to include illumination. While they certainly hurt their attackers, the casualty ratio was not as one-sided on the night of 27 November as one might think. Chinese divisions are not as big as American divisions. On the night in question, the Chinese put roughly four times as many people into harm's way as the Americans,[88] and came away with roughly twice as many casualties.[89] Theirs does not qualify as a suicide attack. Western armies routinely assault with at least a three-to-one edge in men.[90]

## What the Chinese Manuals Hold

As of 1960, four criteria epitomized Chinese offensive action:

a. ... *Four Fast—One Slow.* The *"One Slow"* ... refers to the commander's responsibility for careful evaluation, planning, and inspection prior to the attack. The *"Four Fast"* ... relates to the speed in execution of the attack:
(1) Speed in preparation, including reconnaissance.
(2) Speed in the advance, to flank or encircle the enemy.
(3) Speed in exploitation of gains, to prevent enemy regrouping.
(4) Speed in pursuit, to overtake and destroy a retreating enemy.

b. The *One Point—Two Sides* tactical technique is the launching of a number of separate attacks against one objective ...
(1) *One Point* means to concentrate overwhelmingly superior strength and attack a selected weak point.
(2) *Two Sides* means that when making an attack, two or more efforts of attacking forces are necessary, but it does not mean the attack is limited to only two sides.

c. The isolation and subsequent detailed reduction of individual strong points of a defensive zone are called the *Divide-and-Destroy* tactical method. It is based on

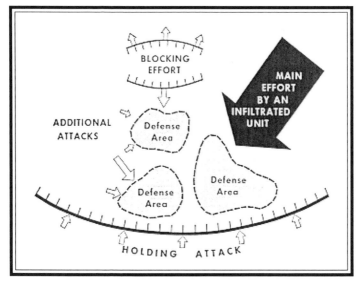

**Figure 7.1: One-Point, Two-Sides Attack on Defensive Position**
(Source: *Handbook on the Chinese Communist Army*, DA Pamphlet 30-51 (1960), p. 24)

the theory that no defensive system can be equally strong everywhere and that weak spots exist which, if captured, will permit an attack from the flank or rear on adjacent strong points.

d. *Strategic penetration* is defined as a massive frontal attack against an enemy in a fixed defense line, the flanks of which are secure. The operation is designed to breach the defense at selected locations, create flanks, and permit passage of mobile forces deep into the enemy rear, to envelop and destroy him.[91]

— *Handbook of the Chinese Communist Army*
DA Pamphlet 30-51, 7 December 1960

## A More Recent Example of the Human Wave

Occasional "mass" assaults also occurred in Vietnam. Corporal Bob OBday, a young Marine machinegunner, observed the same terrifying spectacle from one of the hills protecting the Khe Sanh combat base in 1968.[92] Of note in the North Vietnamese version,

**103**

there was no chanting, yelling, or playing of musical instruments. The NVA had apparently learned from another nation's experiences. It is no coincidence that their fighting method has been characterized as "one slow, four quick."[93]

## Why the Threat of Mass Assault Still Exists

Oriental armies remain capable of successful WWI "human-wave" simulations. Instead of the entire unit on line that Western forces have come to associate with assault, each Eastern wave would consist of squads abreast from different companies. While the lead squads might all attack at once, the subsequent squads would attack whenever they wanted. Instead of the hasty assaults to which U.S. squads must resort when their parent unit's "push" fails, the subsequent squads would conduct "quick deliberate attacks" reminiscent of the WWI German Stormtroopers. These would be attacks that have been prerehearsed as technique and reconnoitered through "recon pull."

The Chinese at the Chosin Reservoir used consecutive squad-sized attacks to force open narrow lanes. This may have constituted a refinement to pre-1950 maneuver warfare methodology. Unlike the German Stormtrooper squads of WWI, each group of Chinamen could watch what happened to the preceding group. Last-minute information on defender strengths and weaknesses gave them more opportunity to choose an appropriate method. Years later, the North Vietnamese further refined the equation by removing much of the obvious signalling, to include pyrotechnics.[94]

The Eastern commander's willingness to shift control to his squad leaders in-mid assault — something unheard of in the West — is what makes this extremely dangerous deception into a marginally effective attack. Each Asian squad leader has several prerehearsed assault techniques from which to choose. He can attack with adjacent squads or by himself. He has the support of the squad behind him. This is quite different from the highly structured and relatively inflexible, Western style of assault — in which whole companies must attack on line and stick to preconceived, and seldom properly reconnoitered or rehearsed, game plans. Oriental armies have enjoyed the benefits of decentralized control for quite some time.

# The Inconspicuous, Low-Land Defense

- *How can Eastern defenders hide in low, flat ground?*
- *What would be their avenues of egress?*

(Source: Courtesy of Cassell PLC, from *Uniforms of the Indo-China and Vietnam Wars*, © 1984 by Blandford Press Ltd.; Corel Gallery, *Clipart — Plant*, 35A032)

## Another Variation on a Familiar Theme

After the Korean War came the trouble in Vietnam. First, the French fell to those whose ancestors had defeated the hordes of Kublai Khan.[1] Then the Americans tried to control territory that had not been successfully occupied since 543 A.D.[2] In theory, flat country favors the more powerful opponent. But at various places across the Vietnamese coastal plain, U.S. forces found themselves fighting for bunkers almost as hard as their fathers had on Iwo Jima. Except this time, water did not restrict the enemy's ability to resupply, reinforce, or evacuate those bunkers. And the newer generation never suspected that the ground had been that well

**105**

prepared. By drawing comparisons across the decades, veterans of the more recent conflict may come to better understand how they could have been beaten while believing themselves to be winning. Contemporary U.S. soldiers should take particular note of the similarities between Chapters 6 and 8. These are not isolated examples, imagined paradoxes, or farfetched hypotheses; these are well-documented trends that U.S. forces will see again. Not all military establishments have difficulty learning from those of dissimilar political ideology.

## Exploiting the Western Mind Set

To capitalize on the "top-down" thought process of a Western attacker, the Oriental defender has only to encourage his individuals and small units to do inconspicuous things. With just two, they could collectively make a difference: (1) covert, spoiling attacks against command posts, supply depots, and supporting arms; and (2) delaying actions on ground lacking strategic value. Then, while forward elements of the attack force may suspect the harm being done by these "little things," their rear-echelon leaders will continue to focus on the "big picture."

While the maneuver warfare precepts have been in the literature since 350 B.C. (with Sun Tzu's *Art of War*), Vietnam era Americans had never heard about this alternative way of fighting either. They were never told how the opposition would try to get into their heads.[3]

## Learning Takes Priority in the East

To grow in proficiency, an infantry unit must squarely confront the tactical outcome and strategic significance of each engagement. When it loses, it must admit that it has lost. In this category, the VC and NVA had the decided edge over their Western adversary.

In January 1969, a twenty-one-year-old main-force [VC] deputy squad leader best explained . . . "Sincerely speaking, I was never disappointed with the result of any battle. Fighting the war we always think that there will be times when we lose and . . . when we win, and we should not be too optimistic when we win or too disappointed when we lose. When we lose we must find out what caused us to lose and gain experience for the next time."4

## The False Face

Again in Vietnam, U.S. forces expended much of their strength against false defensive fronts. The VC/NVA had only to study the manuals of their ally to the north to learn the purpose and specifics of dummy positions. After all, many thousands of "volunteers" from the People's Republic of China were manning the air and ground defenses around Hanoi and Haiphong.5 Some of these volunteers my have helped to train the North Vietnamese Army.

When a structure cannot easily be concealed perfectly, construct many dummy structures so that the enemy will not be able to distinguish the real from the false. These dummy structures will draw enemy fire, disperse enemy fire, and cause them to misuse their forces.

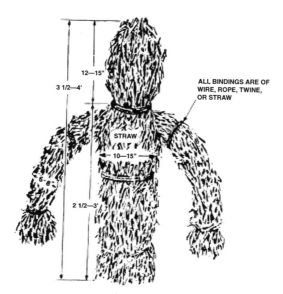

## Figure 8.1: The Visible Defender — "Strawman Decoy"
(Source: Excerpt from *Chinese Manual on Field Fortifications*, in *A Historical Perspective on Light Infantry*, U.S. Army Combat Studies Inst. Research Survey No. 6, p. 72)

The dummy structure should not be too close to the true structure lest it draw enemy fire to the true structure. Moreover, it should not be allowed to fall into enemy hands.

The dummy structure should also be camouflaged and should sometimes be equipped with dummy soldiers and weapons. . . . It is only necessary that it [the dummy structure] agree with the true structure in outward appearance *(Chinese Communist Reference Manual,* 3—4).[6]

— Intelligence Division, U.S. Army Corps of Engineers

## The Teachers' Doctrine

Chinese infantrymen follow 12 general principles of war: (1) strategic goal, (2) security, (3) mobility, (4) local superiority, (5) offensive action, (6) singleness of direction, (7) flexibility, (8) surprise, (9) initiative, (10) unity of command, (11) preparation, and (12) confidence. Most revealing, from the standpoint of defensive action, is the third: "Withdraw before the enemy's advance . . . ; disperse or concentrate one's own forces swiftly."[7]

a. The defensive is assumed in the face of a superior hostile force to permit concentration and employment of troops in a more decisive sector or to lure the enemy into an area that will facilitate offensive operations against him. . . .
b. Chinese Communist defensive strategy stresses mobile warfare based on localized tactical offensive by subordinate echelons. . . .
c. Where applicable, the Chinese Communist concept of defense includes such measures as trading space for time; complete abandonment . . . of areas of little importance; concentration of defensive strength in critical areas; conduct of guerrilla activities in hostile rear areas; and emphasis on psychological warfare.[8]

— *Handbook of the Chinese Communist Army*
DA Pamphlet 30-51, 7 December 1960

The Chinese defender has a different order of work. First, to his front, he establishes "antitank and antipersonnel minefields and obstacles and [fake] field fortifications."[9] Then, he occupies terrain that will itself constitute an obstacle. His fighting holes are camouflaged, mutually supporting, and arrayed in depth. The Chinese defender conceals daytime activity and often withdraws after dark.

The Chinese Communists fight a delaying action when they are at a disadvantage and wish to gain time and wear down the enemy. The overall purpose is to launch a counterassault when the enemy is at a decided disadvantage.[10]

— *Handbook of the Chinese Communist Army*
DA Pamphlet 30-51, 7 December 1960

## Subterranean Factor

Asians construct part of every defense beneath the surface of the earth. The Japanese did it in WWII, the Chinese did it in Korea, and the NVA did it in Vietnam. In Southeast Asia, some of these caves and tunnels were inside the tiny hills and ridges that dotted the coastal plain. Others were so inconspicuous as to require below-water entrances. For further insight, one must refer to the translator's notes to the Chinese manual on field fortifications.

Trench systems . . . are extensive and well-laid-out on the . . . hills. . . . Each hill has one main communication trench following the contour of the reverse slope. From the main trench, short connecting trenches branch off to emplacements and shelters.

The main trench has heavy overhead cover at short intervals; it also has small-arms positions and 1-man shelters cut into its walls. In most cases, the connecting trenches are well-covered; they are tunneled wherever possible, especially between positions on the reverse and forward slopes [of a hill]. . . . All the tunnels are shored with timber. . . .

. . . Individual rifle positions are located on both the forward and reverse slopes for all-round *[sic]* defense . . . [and] interconnected by tunnels. . . .

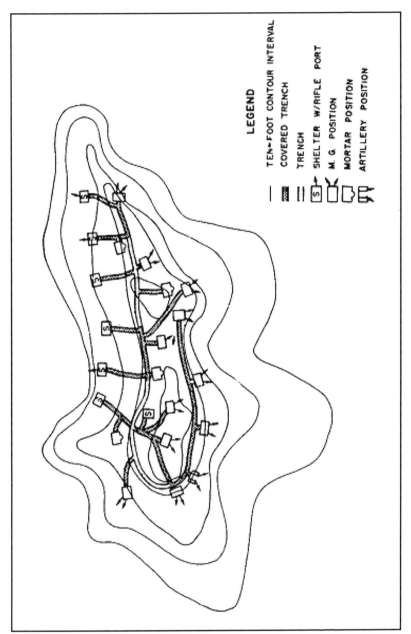

LEGEND

| | TEN-FOOT CONTOUR INTERVAL |
| | COVERED TRENCH |
| | TRENCH |
| | SHELTER W/RIFLE PORT |
| | M. G. POSITION |
| | MORTAR POSITION |
| | ARTILLERY POSITION |

**Figure 8.2: The Underground Portion of Every Defense**

(Source: Excerpt from *Chinese Manual on Field Fortifications*, in *A Historical Perspective on Light Infantry*, U.S. Army Combat Studies Inst. Research Survey No. 6, p. 85)

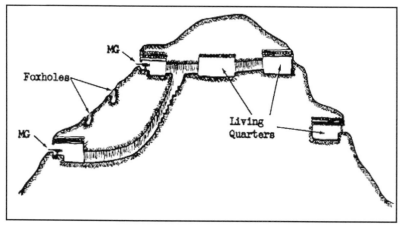

**Figure 8.3: "Machine-Gun Emplacements"**

(Source: Excerpt from *Chinese Manual on Field Fortifications,* in *A Historical Perspective on Light Infantry,* U.S. Army Combat Studies Inst. Research Survey No. 6, p. 88)

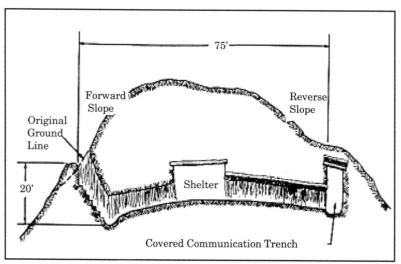

**Figure 8.4: "Tunnel between Forward and Reverse Slope[s]"**

(Source: Excerpt from *Chinese Manual on Field Fortifications,* in *A Historical Perspective on Light Infantry,* U.S. Army Combat Studies Inst. Research Survey No. 6, p. 86)

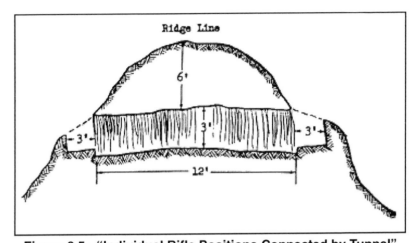

## Figure 8.5: "Individual Rifle Positions Connected by Tunnel"

(Source: Excerpt from *Chinese Manual on Field Fortifications*, in *A Historical Perspective on Light Infantry*, U.S. Army Combat Studies Inst. Research Survey No. 6, p. 86)

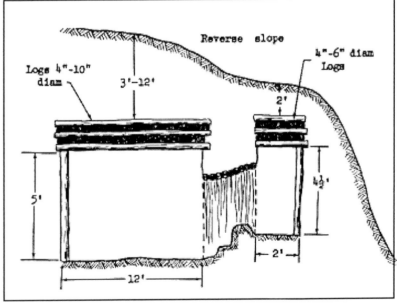

## Figure 8.6: "Cross-Section of Troop Shelter"

(Source: Excerpt from *Chinese Manual on Field Fortifications*, in *A Historical Perspective on Light Infantry*, U.S. Army Combat Studies Inst. Research Survey No. 6, p. 87)

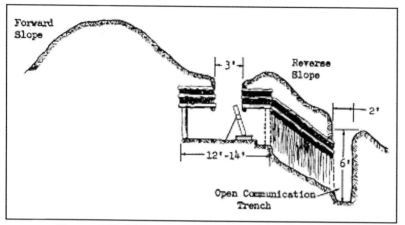

**Figure 8.7: " Mortar Position on Reverse Slope"**
(Source: Excerpt from *Chinese Manual on Field Fortifications*, in *A Historical Perspective on Light Infantry*, U.S. Army Combat Studies Inst. Research Survey No. 6, p. 87)

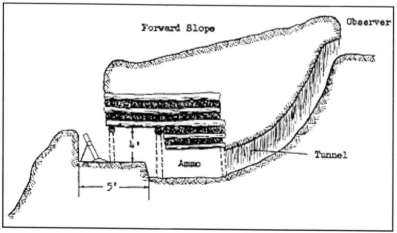

**Figure 8.8: "Mortar Position on Forward Slope"**
(Source: Excerpt from *Chinese Manual on Field Fortifications*, in *A Historical Perspective on Light Infantry*, U.S. Army Combat Studies Inst. Research Survey No. 6, p. 87)

... Troop shelters ... are normally built on reverse slopes. ...

The overhead protection of these shelters ranges in thickness form 3 to 12 feet and consists of many layers of logs and a cover-layer of earth. ...

... Where the terrain permits, mortar emplacements are usually sited on the reverse slopes. ... Most mortar positions are sited to cover dead areas in the field of fire of flat trajectory weapons on the forward slopes. ...

Machine gun ... emplacements are positioned in depth along the forward slopes of hills.[11]

— Intelligence Division, U.S. Army Corps of Engineers

### The Nonexistent Base Camp

Veterans of Vietnam remember all too well a foe who liked to tunnel. How difficult would it have been for him to disguise tunnel entrances or snipe Americans who got too close? Some of these tunnel complexes were too large to be anything but base camps.

To avoid contact, the North Vietnamese and Vietcong [sic] kept out of sight. ... An extensive network of tunnels and underground shelters permitted many local guerrillas to elude Allied patrols. Perhaps the most extensive tunnel system existed sixty kilometers from Saigon in the district of Cu Chi, home to the U.S. 25th Infantry Division. Begun during the French war and expanded during the American war, the complex of tunnels and underground rooms measured perhaps 200 kilometers [124 miles] in all. Extending beneath virtually every village and hamlet in the district, the tunnels contained living rooms and storage areas for food and ammunition connected by passageways. Ventilated by bamboo air shafts, the tunnels were reached by a variety of entrances including camouflaged trap doors and holes dug beneath the water lines of river banks.[12]

### Combining these Factors to Befuddle Western Intelligence

The Vietnamese coastal plain had been intentionally covered

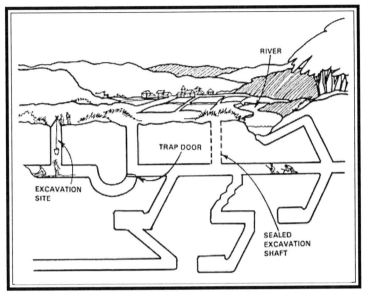

**Figure 8.9: "Underground Base Area"**
(Source: *Counterguerrilla Operations*, FM 90-8 (1986), p. A-5)

with hundreds of thousands of fortifications. That way, no amount of standoff reconnaissance — whether aerial or binocular — could determine which ones were occupied. Some contained hidden tunnel complexes. Those would have allowed VC/NVA units to completely avoid Allied sweeps. In late 1967, a 1st Marine Division reconnaissance patrol working just south of the Cua Viet River mouth witnessed just how easy it was for the enemy to defend.

By 1100 [11:00 A.M. on 27 December], Lima Company was in a world of shit in Than Tham Khe.

Lima had actually been in the village the day before [Map 8.1], sweeping through it at dusk. The presence of trenches and enemy fortifications were noted, but since no special emphasis had been placed on a careful search of the village, the company moved through quickly. . . .

Undiscovered in Than Tham Khe was a heavily armed NVA battalion (the 116th), which had gone to ground the moment Lima Company entered the village. The NVA did not engage the Marines in a fight, and somehow remained hidden until Lima Company passed through. . . .

On the morning of the 27th . . . Lima [Company] set out to return to Than Tham Khe [Map 8.2]. When the company arrived the NVA were waiting for them. They let the lead platoon of Lima walk into their lines before initiating an ambush that inflicted such heavy casualties the platoon was nearly wiped out. . . .

Tragically, the battalion commander . . . started putting pressure on Captain Hubbel [Lima's commander] to get up and take the enemy head on.

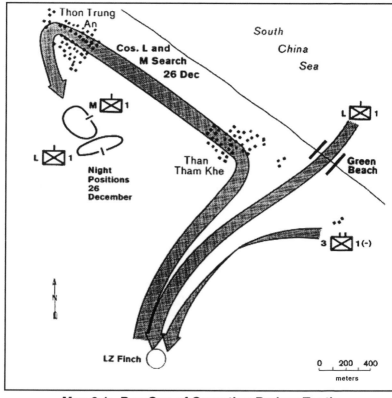

### Map 8.1: Day One of Operation Badger Tooth
(Source: *U.S. Marines in Vietnam: Fighting the North Vietnamese, 1967,* Hist. & Museums Div., HQMC, p. 177)

Captain Hubbel continued to bring down naval gunfire on the enemy. In addition, two separate air strikes were made on Thon Tham Khe. But the NVA were dug in and concealed. . . .

The fire coming from the village did not let up, and every attempt on the part of the Marines to maneuver was met by interlocking fields of fire and mortars. . . .

. . . Captain Hubbel was given what was tantamount to an order to assault the enemy position. . . . [T]he Captain got his men on line, gave the command to charge, and came out of the trench shooting.

. . . They were not on their feet for more than a few seconds. . . . Captain Hubbel and some twenty-six other Marines were killed outright. Those who weren't killed either fell wounded or immediately went to ground to avoid being killed. . . .

At 1130 hours [11:30 A.M.] on the 28th [after a day of bombing and shelling], India and Kilo [with Mike as a base of fire] left their holes and began the assault. Recon was left with the CP [command post] group to watch. The attack went off like a textbook example of an attack against a fortified position. . . .

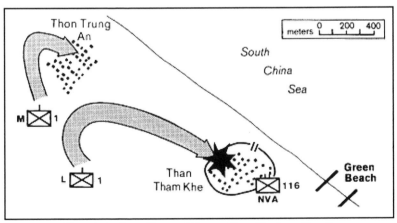

**Map 8.2: Day Two of Operation Badger Tooth**

(Source: *U.S. Marines in Vietnam: Fighting the North Vietnamese, 1967*, Hist. & Museums Div., HQMC, p. 177)

The only problem was, the 116th NVA Battalion had slipped away during the night [Map 8.3], taking all of their dead and wounded with them. India and Kilo swept the objective, taking one wounded but failing to make significant contact with the enemy.

The mystery of where the 116th went and how they got out plagued the battalion for most of the day. . . . Eventually their escape route was discovered in the gap between companies to the west. . . . A synopsis of the battle published in U.S. Marines in Vietnam, 1967 includes a statement that an ARVN [Army of Vietnam] battalion, operating further to the north, discovered over one hundred bodies from the 116th abandoned in the dunes. But I do not recall hearing any such report at the time. . . .

The facts are, that on December 27, 1967, 3rd Battalion, 1st Marines suffered forty-eight dead and eighty-six wounded in the battle for Than Tham Khe, and very few dead or wounded NVA were accounted for in the village.[13]

As Companies I and K made their final assault on Thom Tham Khe during the morning of 28 December, they encountered "heavy

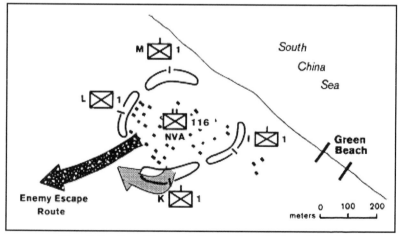

**Map 8.3: Day Three of Operation Badger Tooth**
(Source: *U.S. Marines in Vietnam: Fighting the North Vietnamese, 1967*, Hist. & Museums Div., HQMC, p. 177)

small-arms fire."[14] Apparently, the NVA unit felt it necessary to leave a rear guard. A detailed search of the village revealed some clues to what had happened.

> This search revealed a village that was literally a defensive bastion. It was prepared for all-around defense in depth with a network of underground tunnels you could stand up in, running the full length of the village. Connecting tunnels ran east and west. This tunnel system supported ground level bunkers for machineguns, RPGs [rifle-propelled grenades], and small arms around the entire perimeter of the village. Thus the NVA were able to defend, reinforce, or withdraw in any direction. All defensive preparation had been artfully camouflaged with growing vegetation. Residents of Tham Ke [sic], questioned after the fight, disclosed that the NVA had been preparing the defense of this village for one year (Col. Max McQuown, Comments on draft, 20 May 1981, MCHC, Washington, D.C.).[15]
>
> — Battalion Commander, 3rd Battalion, 1st Marines
> *U.S. Marines ... Fighting the North Vietnamese, 1967*
> History and Museums Division, HQMC

How the NVA battalion got away is not precisely known. It may have slipped through the U.S. cordon or merely rehidden below ground. That it suffered 100 casualties in the process is doubtful. But whatever happened must have severely demoralized the U.S. unit. This is, after all, the objective of maneuver warfare.

## The Large-Unit or "Multiple-Line" Defensive Array

VC and NVA defenses didn't fit the U.S. pattern. They were designed to weather air and artillery attack. Instead of perimeters, they consisted of roughly parallel lines of fighting holes.

> Under the dense canopy of vegetation, two lines or belts of fortifications were constructed fifty to two hundred meters apart. . . . These belts of defensive positions followed the outline of an L, U, or V so as to offer the possibility of a crossfire. . . .

The distance between the first and second lines of fortifications was dictated primarily by the consideration that the second line not be visible from the first. This second line of defense provided the defenders a protected position to which to withdraw if pushed out of the first belt. From there they could further withdraw from the contact area of counterattack to regain the first belt. . . .

Another deception method was to construct dummy positions to draw artillery and air attacks. . . . Base camps followed the same two-belt design as the temporary bivouac sites with two primary differences — a third belt was added in the larger camps and fighting positions/bunkers were larger and better protected. . . .

One tactic developed by the VC/NVA to counter this "surround and pound with fire support" technique was to "hug the enemy" rather than fall back to the second belt of defenses. This involved following the attackers as they pulled back to put in artillery and air.[16]

## The Small-Unit or "Wagon Wheel" Defensive Array

As was discussed in Chapter 4, the Asian has a fondness for the V-shaped formation. It provides him with a ready-made firesack into which his pursuers can wander. When the firing starts, those pursuers have nowhere to hide. Also observed in Vietnam was a small-unit defensive array that utilized a combination of "V's." It was shaped like the spokes of a wheel. With the fields of fire carefully marked, each element can cover the fronts of the other two.

## The Strongpoint or "Starburst" Defensive Array

Almost every diagram of the Japanese defenses in the Pacific reveals a nonsensical pattern of trenches, individual positions, and clusters of bunkers.[17] The circular clusters may have been the semi-independent fortresses in a matrix of strongpoints. The same thing occurred in Vietnam. And, as had happened in WWII (see Figure 6.2), the bunkers in each circular cluster were connected to a central bombproof shelter. Like the "Wagon Wheel," the resulting formation incorporated many firesacks.

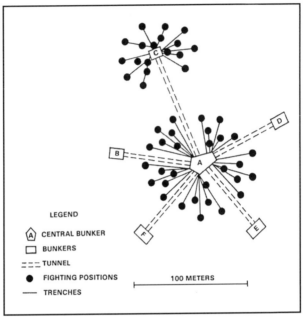

**Figure 8.10: The Strongpoint Defensive Array**

(Source: *Counterguerrilla Operations*, FM 90-8 (1986), p. A-6)

## The Subterranean Withdrawal

After paying the price to enter a contested position, American troops routinely discovered bunkers, trenches, and too few enemy bodies. One wonders how many defenders may have escaped through tunnels or waited below ground for the U.S. forces to leave. After the final battle of the second Tet offensive, large quantities of rice and weapons were found below ground at "La Thap (1)."[18] A subterranean avenue of egress may have been responsible for the low enemy body count.

## The Sapper-Conducted Spoiling Attack

Many of Iwo Jima's "coincidences" also happened in Vietnam. They were unmistakably the work of sappers.

In Vietnam, U.S. tanks had difficulty participating in the big battles. Before they could enter the fray, they would invariably hit a land mine. So many planes, ammunition dumps, and other targets of strategic significance were lost to indirect enemy fire that each occurrence would be difficult to document. Suffice it to say that the volume of fire was so much lower in Vietnam that the "direct hits" more clearly violated the laws of probability.

## The Next Time Around

The lessons for the contemporary U.S. soldier are sobering. Having a heritage of battles that cost more than they accomplish, he must carefully weigh the strategic value of each attack. Past adversaries (Germans, Japanese, North Koreans, Chinese, and North Vietnamese) have all been adept at small-scale maneuver warfare. Other potential adversaries (former Soviet client states) know how in theory to defend a piece of ground without getting hurt.

(Source: *FM 21-76* (1957), p. 71)

U.S. maneuver elements of the future — no matter how well supported by armor, indirect fire, and airpower — should not expect to find contested ground of little strategic value still occupied. Nor should they be surprised by being fired upon from areas already secured. Nor should they be disturbed by having no one to shoot back at. Their quarry has several ways in which to disappear: (1) hide below ground, (2) use a secret escape tunnel, and (3) exfiltrate U.S. lines.

American units should expect their attack lanes and objectives to be defended from other locations — by long-range machinegun fire from indestructible positions to the left or right. They should also expect sniper or machinegun fire from the rear.

# The Absent
# Ambush

● *How can a U.S. patrol find itself instantly surrounded?*

● *In what way does the Eastern unit finish off its quarry?*

(Source: Courtesy of Cassell PLC, from *World Army Uniforms since 1939*, © 1975, 1980, 1981, 1983 by Blandford Press Ltd.; Corel Gallery, *Clipart* — Plants, 35A007)

## The Close-Range-Combat Differential

In Vietnam, U.S. infantrymen experienced intermittent periods of "monotony" and "ambush." They would stand perimeter watch and run security patrols with very little action for weeks. Then, on some unit operation, they would come under withering fire for a matter of minutes. Why, for them, did warfare fluctuate so wildly between its extremes? Why weren't there more times when the enemy got surprised? The explanation is quite simple really — only difficult for Americans to accept. An adversary of superior individual or small-unit skill will only be found when he wants to be. Every contact with him will seem like an ambush.

Shocked by the encounter, his adversaries can only react (as opposed to seizing the initiative). Forced to react continually, those adversaries develop a defensive mentality.

Fully aware of this "Achilles' heel" in America's military preparedness, Eastern foes have frequently taken advantage of it. With longer traditions of maneuver warfare, they have allowed their lowest echelons to exercise more initiative. To escape U.S. supporting arms, they have withheld their fire until pursuers were almost upon them. To keep U.S. units off balance, they have routinely used deception to build surprise, and surprise to build momentum.

For the maneuver warrior, the distinction between offense, defense, and ambush blurs. For him, ambushes in series constitute a type of defense. Often, what happens after stumbling upon a deliberate ambush or defensive fort in the Orient is the only way to distinguish between them. In both cases, the occupants will do almost the exact opposite of what Western soldiers would do under similar circumstances. In this way, the Easterners lower the chance that their actions will be anticipated.

From a prepared defensive position, the Asian will sap his attacker's momentum and then withdraw. He has replaced the *banzai* mentality of a maneuver warfare novice, with something less predictable. He often only counterattacks locally to reacquire something of strategic value — prominent terrain or an opposition asset. But this counterstroke may be delayed or unrecognizable.

From a prepared ambush site, the Asian will thoroughly disorient his quarry and then close in for the kill. He will abandon good cover to do so. With less wherewithal and more small-unit ability, he prefers to personally administer the killing blow.

### Minimizing the Response from U.S. Supporting Arms

To escape U.S. bombardment, Oriental ambushers would have to stay close to, intermingle with, and then secretly depart from their victims. To do so safely, they must kill or capture all of those victims. They would need a particularly lethal trap. The "U" or "V"shape ambush qualifies. After luring the Western unit into the center of the "U," the Easterners could hit it with precision, long-range fire from the front, and well-aimed, close-range fire from the flanks. Then, they could close the top of the "U" and tighten the noose.

For bait, the Asian might use scouts unwilling to make contact or a sniper hesitant to take well-aimed shots. After the Western unit had deployed to chase the rabbit, its commander might get sniped and its main body subjected to intense machinegun fire. Then comes the tightening of the circle. Isolated pockets of resistance would be easier to handle, so the strung-out victim must be dissected somehow. All the while, minefields, mortar bombardments, and snipers would discourage any reinforcement of the beleaguered unit.

With its cohesion gone, the Western unit could be eventually destroyed piecemeal. Its commander would detect too late that his troops had been lured into the center of an enemy formation, encouraged to deploy against one side, and then mowed down from all sides. Throughout the engagement, his tormentors had been too close to engage with supporting arms. He might even see the folly of conditioning U.S. ground troops to hastily assault any ambush within 50 yards.

## What this Type of Ambush Might Accomplish

One of the best-documented ambushes of this variety involved Companies A and B of 1st Battalion, 9th Marines while they were operating just northeast of Con Thien in early July 1967. Because it occurred along a major north-south infiltration route between routine American sweeps of the area,[1] it may have been hastily arranged by an NVA unit trying to sneak across the border. Enemy scouts had been watching Company A for many hours,[2] so this was no "chance contact."

The ambush occurred just north of the McNamara Line — a 200-meter-wide trace paralleling the Demilitarized Zone (DMZ). To be precise, it occurred along Route 561 — a narrow, north-south cart trail that intersected the trace 1200 yards east of Con Thien. The initial contact was made by Company B in what Marines called the "Marketplace" — an area of widely dispersed concrete ruins. It occurred between two trail junctions that were 1200 and 1700 meters north of the trace, respectively. Several hundred meters north of this "kill zone" was a low, east-west ridgeline. A short distance to the southeast was an elephant-grass-covered rise that has come to be known as Hill 70. Route 561 was, for all practical purposes, a relatively straight, sunken road.

**127**

The trail was eight to ten feet wide and bordered by hedgerows that grew along the trail's embankments and were actually above the helmets bobbing on the road, especially where the trail was worn three or four feet below ground level.[3]

To the right and left of the Route 561 were treelines, dried-up rice paddies, clumps of bamboo, and occasional ruins. Old bomb craters, fighting holes, and trenches were also liberally sprinkled across the area. It was Company B that first realized it was in trouble. At the time (9:30 A.M. on 2 July 1968), Company A was 1500 meters to the west and also moving north.

### How this Technique Was Applied at the Marketplace

The ambush that would initiate Operation Buffalo had several disturbing refinements to what might normally be expected of a maneuver warfighter. The right and left sides of this "inverted U" had different missions and methods.

First, from the northwest, a single sniper opened up. Then, some NVA in a trench 100 meters to the west of Route 561 started taking pot shots at the Marines. They must have known that U.S. troops think in terms of "contact right" and "contact left," but seldom "contact right and left." They must have further known that American infantrymen will hastily assault a small contingent of "Third-World" adversaries. True to form, the Marines assaulted and got cut to shreds by the theretofore-quiet machinegunners in the trench. Then, to the utter dismay of the assault survivors, the entire NVA force on the left began to displace toward Route 561. It did so not as a human wave, but rather as a ragged line of occasionally moving "grass clumps."

Meanwhile, the road had been under intermittent mortar and machinegun fire from the north. And NVA soldiers manning a belt of mutually supporting spider holes along the right side of the inverted "U" had started picking off the exposed Marines from the rear.

Then, a third NVA force moved down from Hill 70 (behind the Marines) to close off the bottom of the inverted U. And elements of the western contingent of NVA crossed the road in several places to split Company B.

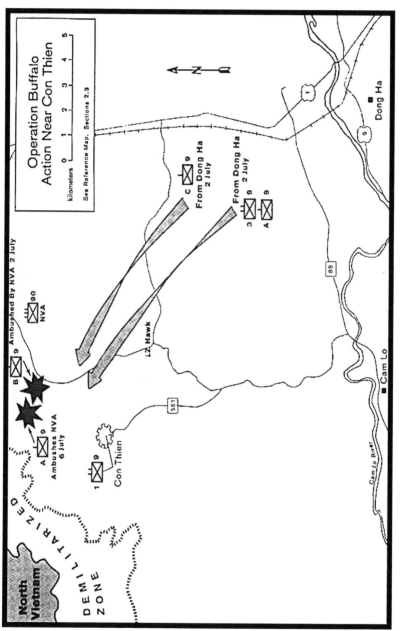

**Map 9.1: Operation Buffalo**

(Source: *U.S. Marines in Vietnam: Fighting the North Vietnamese in 1967*, Hist. & Museums Div.,HQMC, p. 97)

## Eliminating the Opposition Command Group

The first step in a maneuver warfare ambush is often to disrupt the quarry's command-and-control apparatus. This is most easily accomplished by eliminating leaders or radio operators.

In the ambush along Route 561, the first thing that happened was no coincidence. The commander of the lead platoon got wounded only badly enough to get removed from the action.

> The firing on Bravo Company had begun beyond the second crossroad on Route 561, the leftward leaning T [shaped] junction. . . . The 3rd Platoon's first casualty was Staff Sergeant Reyes, their platoon commander, his trigger finger was shot off by one the first sniper shots. He started back down the column to be medevacked as his platoon wheeled to the left flank (west) to assault what appeared to be a squad of snipers in a trench of some sort.[4]

Before what had promised to be a minor skirmish was to end, most of the Company B's officers, staff NCOs, and radio operators would be dead. Their willingness to lead by example had been only partially responsible for their demise.

> This fire killed Captain Coates [the company commander], his radio operator, two platoon commanders, and the artillery forward observer. The attached forward air controller Captain Warren O. Keneipp, Jr., took command of the company, but soon lost radio contact with the platoons [and later his life]. Only the company executive officer, at the rear of the 2nd Platoon, managed to maintain contact with the battalion CP, but the heavy fire [and heat prostration] kept him from influencing the situation.[5]
> — *U.S. Marines . . . Fighting the North Vietnamese, 1967*
> History and Museums Division, HQMC

## The Bait

The maneuver warfighter would make every effort to draw his quarry into a well-laid trap. He could do so by fielding soldiers who run away or shoot from a distance.

On the day before the ambush at the Marketplace, a pair of NVA scouts had wandered down the center of Route 561,[6] and a sniper had taken a single rifle shot at the Marines' nighttime bivouac.[7] When 3rd platoon received sniper fire from the northwest on 2 July 1968, it deployed and moved in that direction. When it got a little more fire, it assaulted. This was all part of the setup. No one had seen the need to discuss the finer points of Oriental tactics with low-ranking U.S. infantrymen.

> They [the 3rd Platoon] assaulted right into a *platoon's* worth of AK47 fire. . . .
>
> . . . The number of AK-firing NVA multiplied again — they just seemed to materialize inside spiderholes and bunkers in the elephant grass and hedgerows — and the entire 3rd Platoon was soon engulfed. The NVA were utilizing those pre-dug ambush positions that invisibly dotted the Con Thien countryside. And as experience had taught them, they had held the bulk of their fire until their foes were within a few dozen meters, so that the soon-to-be-pinned-down Marines could not employ air or arty support without endangering themselves. From their concealed position, the NVA trained their AK47s down prepared fields of fire, squeezing point-blank bursts into confused Marines seeking cover and brave Marines desperately rising up to find something to shoot back at. Noise and chaos and screams, and Corporal Bell [acting platoon leader] no longer answered his radio.[8]

When the Bravo Company Commander realized what had happened, he immediately took action. Sadly, it was too little and too late.

> The NVA had originally been firing [single-shot] SKS carbines, but when AK47 automatic weapons joined in, Captain Coates radioed Staff Sergeant Burns of 1st Platoon. Burns [several hundred yards to the south along Route 561] was ordered to load one of his squads with M60 ammunition and rush them and a machinegun team up Route 561 to reinforce the contact, and have the other two squads standby *[sic]* as needed.[9]

### The Sunken Road Becomes a Kill Zone

At the top of the inverted "U" would be a machinegunner who could fire down the long axis of the opposition column. With the "direct-lay" or line-of-sight method, a mortarman could also shoot accurately from that location. To blanket a relatively straight stretch of road with mortar rounds, he has only to spin the elevation wheel.

The ambush at the Marketplace was sprung so that a bend in Route 561 would mark the top of the inverted U. The Marines were forced to take cover in the relatively straight portion of trail to the south — i.e., a "danger zone" in terms of frontal fire. To make matters worse, NVA mortar and artillery observers had undoubtedly preregistered both junctions.[10] They could place accurate fire all along this portion of the trail without hitting their brethren on either side.

> The number of wounded and dead mounted as NVA fire hit the unit from the front and both flanks. To worsen matters, the enemy began pounding the Marines with artillery and mortars.[11]
> — *U.S. Marines ... Fighting the North Vietnamese, 1967*
> History and Museums Division, HQMC

### The Encirclement

An ambitious ambusher would block the open end of the "U" to keep his quarry from being withdrawn, reinforced, or resupplied.

At the Marketplace, the North Vietnamese did just that. Before the Marines ever realized that they had opposition to the right, an NVA unit was moving down from Hill 70 to shut the back door.

> Fifteen minutes after Captain Coates, Bravo Six, had instructed Bravo One's Staff Sergeant Burns to send a squad and a machinegun team to reinforce the contact, Lieutenant Delaney, Bravo Five, ordered the other two squads forward. . . . [They] were moving up Route 561 when the point squad leader, Corporal McGrath, called to Staff Sergeant Burns that they had [enemy] movement on their right flank on Hill 70. . . .

The NVA kept coming amid the explosions [of Marine mortar shells] in fireteam-sized rushes, and Lieutenant King, Bravo Two, decided to tighten up their perimeter along the road. Getting a hold of Sergeant Hilliard through the 81mm FO's [forward observer's] radioman, he told him to pull Lance Corporal Herbert's point squad [scouting east of the road] back with him.[12]

## The Right Side of the Trap Appears

The U-shaped ambush would work better if the detachment forming one leg stayed hidden and delayed shooting. While under heavy fire from the front, the victims would be less likely to notice fire from the rear.

While 3rd Platoon was becoming decisively engaged to the west of Route 561, 2nd Platoon was looking east of the road for a way to outflank the enemy.[13] When they encountered fire from every direction, they pulled back to the road. They had stumbled into a carefully designed matrix of mutually supporting spider holes. To make matters worse, the hole occupants were taking turns shooting. This made it hard for the Marines to accurately return fire.

After having his platoon sergeant, radioman, corpsman, and point man all killed in the first moments of the attack on 2nd Platoon, [squad leader] LCpl Hebert scrambled into a small bomb crater where three other Marines had also sought shelter.[14]

## The Hammer and Anvil

The maneuver warfighter would close with his opponent to escape bombardment. He values proficiency at close combat more than his attritionist counterpart. In a U-shaped ambush, he might send one side in against the stationary other.

When more firing erupted from the southern portion of the treeline west of the road, Captain Coates did what the NVA knew he would. He did what he had been trained to do — further develop the situation. Unfortunately, by advancing against a superior opponent, he only managed to make matters worse.

NVA fire continued to snap by, mostly from a treeline across a fairly barren field to their left [west] at about the eleven o'clock position. . . . When the fire slackened a bit, Captain Coates (who was walking up and down Route 561 and occasionally firing his .45 into the brush) shouted at Haines to get his [reserve] 2nd Platoon squad on line with the first squad that had come up from 1st Platoon. Together, they were to assault the enemy-held treeline at eleven o'clock [to the west]. . . .

. . . [Soon] any man who exposed his position in the brush by firing, instantly had that NVA machinegun blasting at him. The NVA were dug in and could see everything. The Marines did not have a single target to square their sights on. Nothing in the NVA treeline could be seen moving. . . .

The Marines' fire petered out. A sergeant yelled back that the NVA fire was too heavy to advance any farther and Captain Coates gave the word to pull back. Too late. The two squads were pinned down.

Now the NVA tried to surround them. . . . Hendry looked up and there the little bastards were, streaking through the brush and on across the road to take up positions about a hundred meters north of their pinned-down situation.[15]

In essence, the unit on the left side of the inverted "U" had assaulted the road. Already on line at the front of the treeline just to the west of Route 561, it had only to move forward in the same formation. But this was not a Western-style, perfectly aligned assault. Its participants were well enough trained to displace by fire and movement (in three-man fire team rushes). The NVA soldiers had camouflaged themselves to look like grass clusters. Every time one of the fire teams stopped, it blended in so perfectly with the savannah that the U.S. troops had nothing to shoot at. These were opponents with more skill than the Marines had believed possible.

They opened fire on the NVA still dashing across the road. They fired on the NVA in the brush on their flanks too, but with elephant grass secured to their pith helmets and bush hats as well as on their packs and web gear, and around their arms and legs, the NVA were almost impossible to see.[16]

## Dissecting the Victim

To annihilate an entire column of enemy troops, one must subdivide it — interrupt its sharing of information and assets. When one side of a U-shaped ambush closed against the other, it might at the last moment send out narrow thrusts to segment its strung-out victim.

At the Marketplace, elements of the NVA unit assaulting from the west crossed the road. They must have been trying to dismember Company B. By mixing with the Marines, they would totally escape U.S. supporting arms.

## The Kill

Maneuver warfighters care more about destroying targets of strategic significance than killing enemy soldiers. However, there are times where proficiently dispatching one's opponent is the only way to safely occupy his space.

The NVA at the Marketplace displayed some highly unusual and lethal ways of removing opposition. Most effective were the random shots from the spider holes to the east of the road. Those shots must have been impossible to hear above the western cacophony. By only occasionally firing, the spider hole occupants kept the Marines confused as to their precise locations. Their comrades west of the road employed other tricks. Some shifted location between shots to hide their whereabouts, and others to get a better view.

> There were NVA climbing up in the trees to fire down on the pinned-down Marines from 1st Platoon and the pinned-down Marines from 2nd Platoon [to the west of Route 561].[17]

The *coup de grace* was complete when the Bravo Company CP was occupied.

> They ended the stalemate by overrunning the Command Post and the 2nd Platoon of Bravo Company. . . .
>
> The main body of NVA reached the intersection of Route 561 and the east-to-west trail in a screaming, shooting fury . . . and Captain Keneipp, only days away from reassign-

ment to the astronaut training program at NASA, provided Bravo Six's last words to the CP at Con Thien . . . "I don't think I'll be talking to you again. We are being overrun."[18]

## How Reinforcement Was Prevented

To lower the possibility of counterattack, the maneuver warfighter isolates the battlefield.

When Company B got ambushed that day in early July 1967, its sister company tried to come to its assistance immediately but could not. Operating 1500 meters to the west, Company A had its own problems. They all started when several Marines tripped some recently installed booby traps.[19]

> Power found the trip wire of one of the booby traps, or rather he found the end that had survived the explosion. *Brand spanking f____ new,* he thought as he inspected the wire. *Ain't roughened, ain't corroded, ain't weathered, ain't nothin.* The NVA had probably rigged it [the booby trap] last night or this morning. . . . *I know these m_____ are looking at me right now. They are right up here.*[20]

Then Company A's medevac chopper got mortared,[21] and its CP received 30-40 more rounds within three minutes.[22] As the crescendo from the east intensified, Company A was ordered to move along the road that led into Route 561's southern junction.[23] But the North Vietnamese had other ways to prevent Company B's reinforcement.

> As Alpha Company hurried through the waist-high brush, trying to close that 1,500 meter gap with Bravo Company, NVA snipers fired on them from concealed positions to the north and east, though not from the . . . south and the closer they got to Bravo, the heavier the fire on Alpha. These AK- and SKS-toting snipers, all poor shots, did not slow Alpha Company down, but the claymore mine that the point team walked into did.[24]

Then the machineguns that had taken Company B under long-range fire opened up on Company A.

[T]he [Company A] grunts could hear a considerable number of NVA heavy machineguns open fire on them from a tree-lined ridge far to the north. The NVA were swarming all around Bravo company but also had units all the way north to that ridgeline.[25]

By now, Company A had taken so many casualties that Captain Slater could only send one platoon to help Company B.[26] While his medevac LZ (landing zone) later came under ground assault,[27] that single platoon made it over to Route 561 and managed to survive the final onslaught.[28]

An armored relief force came up the trace from Con Thien but ran into trouble before it could reach Company B. When Company C helicoptered in from Dong Ha, it came under enemy artillery fire at its LZ on the trace.

A North Vietnamese unit, trying to encircle Company B, had moved south and was opposite Radcliffe's small force. Helicopter gunships and the fire from the four tanks dispersed the enemy. Company C began arriving by helicopter and Captain Radcliffe ordered the Company D platoon to secure the landing zone and evacuate casualties. As the lead elements of Company C came into the zone, they met a heavy artillery barrage, which wounded 11 Marines.[29]

> — *U.S. Marines ... Fighting the North Vietnamese, 1967*
> History and Museums Division, HQMC

(Source: *FM 21-76* (1957), p. 70)

## Deja Vu

This stage of Operation Buffalo was over, but not before some pretty good Marines had to learn "the hard way" what historians could and their headquarters should have told them. The NVA formation at the Marketplace was almost identical to the *Haichi Shiki* of Chapter 4. Sadly, the young Leathernecks did not know what their predecessors had seen in Korea. Nor were they aware of what the U.S. Army had encountered at Ia Drang in the autumn of 1965 — a "Flexible Horseshoe."[30] All the while, the Chinese manuals clearly described the maneuver.

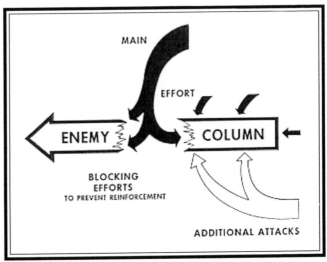

**Figure 9.1: One-Point, Two-Sides Attack on Enemy Column**
(Source: *Handbook on the Chinese Communist Army*, DA Pam 30-51 (1960), p. 24)

Had the young Americans expected a trap, they would not have taken the bait. They could have scouted the snipers' location and discovered it to be a prepared enemy position. Once the extent and shape of the enemy formation had been determined, the Marines could have countermarched to the south. If the road had come under long-range machinegun and artillery fire, they could have moved to the west a short distance "on line" and then south in column.

# 10 The Transparent Approach March

- *How do Easterners reach their attack objectives unseen?*
- *In which ways can they defeat U.S. Intelligence?*

(Source: Courtesy of Cassell PLC, from *World Army Uniforms since 1939*, © 1975, 1980, 1981, 1983 by Blandford Press Ltd.; Corel Gallery, *Clipart — Plants*, 35A019)

## The Easterner's Favorite Illusion

The Asian soldier is a master of the approach march. His tradition is to attack out of nowhere — to suddenly appear where least expected. In the fall of 1950, several Chinese divisions managed to sneak across the Manchurian border, attack U.S. units around the Chosin Reservoir, and then disappear into the blowing snow. Their secret had been simple — mobility and camouflage.

> Although one is near, give the impression that one's forces are far away.[1]
> — *100 Strategies of War*

## The Asian First Enlists the Help of the Local Populace

In Vietnam, North Vietnamese assault forces often approached their objectives through long-range infiltration. Their subordinate elements entered the coastal plain along different routes and then reassembled near the parent unit's objective. To transit the many miles of U.S. free-fire zones more quickly, those elements obtained much of their ammunition and supplies from caches along the way. One wonders to what extent those routes and caches may have been available to any unit in transit. It is known that main-force Viet Cong assault units remained constantly on the move to avoid Allied counterattack. To do so, they would have needed prearranged resupply points.

> The objective of the VC/NVA in avoiding contact with ARVN and U.S. forces was achieved by constant mobility and elaborate information-denial procedures *(Insurgent Organization and Operations,* RAND Corporation Study RM-5239-1-ISA/ARPA, pp. 59-101). . . .
> To accomplish this requirement of constant mobility, the Front units often moved daily and never stayed in one location more than four days. . . .
> Camp sites had to meet three main criteria. If at all possible, they were to be established under dense foliage to deny observation from the air. The distance between camp-sites had to be a single night's march or less. Finally, the sites had to have some minimum potential for defensive works.[2]

Contrary to popular opinion, the Viet Cong shouldered most of the responsibility for security and reconnaissance within their respective locales.[3] They could support North Vietnamese maneuver forces in a number of ways: (1) identify promising targets, (2) reconnoiter objectives from the inside, (3) guard infiltration routes and supply caches, (4) follow U.S. ambush patrols into position and mark occupied trails, and (5) guide assault elements through the protective barriers of mines, barbed wire, and early warning devices.

> An incoming intelligence report was the start of most main-force Viet Cong [or NVA] operations. The usual source

of intelligence was the local guerrillas, Viet Cong part-timers who worked the land for a living but took time off to plant booby traps, carry out raids on weakly defended targets or reconnoitre US firebases and LZs. Reports would come in to the headquarters of a main-force Viet Cong [or NVA] regiment, indicating a number of potential targets. If the regimental commander liked the sound of one of these, he would send out some of his own reconnaissance personnel to contact the Viet Cong villagers and be taken on a guided tour of the objective. Then, if everything seemed right — the avenues of approach, assault and withdrawal all good — the operation would be authorized, a unit assigned and detailed planning begun.

Soon every man in the chosen [main-force] Viet Cong [or NVA] unit would know the target like the back of his hand — every defensive installation, building, fuel store, weapon emplacement. And he would know exactly what he personally had to do, the route of advance, his part in the assault, the assembly point after the fight and the various routes back to base. . . .

The objective would often be several days march from the [main-force] Viet Cong [or NVA] base in the Highlands, or across the border from Laos or Cambodia. To remain unobserved during this advance was essential. The small columns of men would thread their way silently through the jungle, dropping to the ground at the sound of an aircraft. They carried twigs and leaves attached to wire frames on their backs that provided perfect camouflage once the men were flat on the ground.

Once out into populated farmlands, the [main-force] guerrilla [or NVA] columns marched by night, guided by local guerrillas. In this way the [main-force] Viet Cong [or NVA] could normally assemble a complete battalion or more near its target without the enemy suspecting their presence. . . .

The attack itself would began [sic] after dark. . . . Again aided by the local guerrillas, the assault force would move up to positions just outside the enemy's defensive perimeter. At zero-hour, a barrage of mortar and rocket fire would pound the target and then guerrillas would storm forward, pressing the fight to close quarters. . . .

**141**

... But as soon as it was felt that the tactical objectives had been achieved, the order would come to withdraw. The aim was to deny the Americans time to react, and not to be caught by a counterblow.

Clearing the battlefield was an important part of any operation. The Viet Cong [or NVA] were quite prepared to risk lives to retrieve dead bodies for proper burial. . . . The weapons abandoned by their own dead or wounded and any enemy arms were gathered up, as other soldiers maintained covering fire. As the withdrawal began, a rear guard took up position to deter any pursuit. . . .

The assembly point for the retreating force was usually set about 12 hours march from the scene of the action. . . . Local village guerrillas were always essential for guiding main-force units during a hurried withdrawal in a populated area, because only they knew how to avoid all the booby traps that littered the pathways. They also had the latest information on enemy patrols and could steer the soldiers clear of potential encounters which had to be avoided at all costs.[4]


## Approach March Preparations

The NVA reconnoitered their attack routes ahead of time. In the process, they gathered intelligence from local residents and arranged for local guides. For the Asian, human intelligence takes precedence over the signal or aerial variety.

Movement methods followed the same general procedures regardless of the size of the unit. Although movement doctrine was designed for battalion-size forces, it was easily adapted to units as small as a squad. . . . The day prior to the move, [several] reconnaissance elements from the moving unit . . . made contact with district or village cadre to arrange guides and provisions and to familiarize themselves with the movement route.[5]

It is doubtful that North Vietnamese maneuver elements prestaged their own supplies before each attack. More probably, the Viet Cong maintained a network of prestocked way stations.

They may have even kept watch over the routes in between. During the many "search and destroy" sweeps in Vietnam, U.S. rifle companies often stumbled upon buried caches of ammunition, weapons, rice, and medical supplies. The sound of gunfire or sight of discarded U.S. gear (signifying a past firefight) often preceded these discoveries.6 The buried supplies and equipment were obviously being guarded by someone.

## The Use of Established Infiltration Routes

In Southeast Asia, certain sectors of each U.S. Tactical Area of Responsibility (TAOR) generated more firefights than others. While few U.S. commanders realized it at the time, those sectors probably contained established infiltration routes — covered trails between well-supplied way stations. If disturbed at one of the smaller rest stops, a transiting unit would fight. In early 1967 at the "Three Gateways to Hell" portion of the trail between Gio Linh and Con Thien, Delta Company, 1st Battalion, 4th Marines found that out the hard way. Little did they know that 100 yards away lay the fully camouflaged entrance to a tree-covered, cobblestone road running north into the DMZ. The dug-in enemy battalion easily fended off a hasty platoon-sized flanking attack and then waited for evening. When Alpha and Delta companies assaulted the location the next morning, they found a relatively straight string of fighting holes connected by land lines, but no bodies.7 This was no fluke. The same kind of thing was happening all over Vietnam.

After paying the price to enter way stations that had been totally surrounded, Americans would routinely find fortified emplacements, and too few enemy bodies. One wonders how many defenders may have escaped through tunnels or waited below ground for the U.S. units to leave. The NVA routinely disguised buried caches of ordnance and supplies, why not the entrance to subterranean avenues of egress?

Clearly marked on every U.S. intelligence officer's map were all the enemy sightings, cache retrievals, and firefights that had occurred within his TAOR. To locate the opposition's avenues of approach to the big bases, he had only to connect the dots like the spokes on a wheel. An in-depth terrain study would have further revealed the enemy's probable path between dots. But such conclusions were not drawn during the Vietnam War. The obvious infiltration routes were not exploited. By simply backtracking along them, U.S. forces could have found the major arteries to the Ho Chi Minh Trail.

This revelation did not become a "lesson learned" or enter the corporate memory of the U.S. military, because the way stations were so perfectly camouflaged as to hide their true purpose. What if they had routinely sported a sizeable, but well-hidden, underground room? Could not that room have doubled as bomb shelter and additional escape option for rear-guard defenders?

## Along those Routes May Have Been "Forward Base Areas"

Some of these infiltration routes may have led through the center of what Mao Tse-tung referred to as "temporary base areas."

Some of the semipermanent stations nearer population centers or in areas of frequent Allied sweeps were underground in elaborate tunnel systems.[8]

By 1945, Mao had concluded that "base areas" were prerequisite to guerrilla warfare,[9] and that guerrilla warfare could evolve into mobile warfare.[10] In Vietnam, American planners talked of base areas only in remote or mountainous regions. But Mao Tse-tung made it clear that base areas could also be established on broad plains.

As to the possibility of establishing on the plains base areas that can hold out for a long time, it is not yet confirmed; but the establishment of temporary base areas has been proved possible, and that of base areas for small unit or for seasonal use ought to be possible. . . .
. . . [I]t is not impossible for numerous small guerrilla units to scatter themselves in numerous counties across the broad plains and adopt a fluid mode of fighting, i.e. to shift their base areas from one place to another.[11]
— Mao Tse-tung

There was reputedly an area of heavy tunneling just 3000 meters to the west of An Hoa, where the Song Thu Bon River enters the coastal plain.[12]

## The Final Rest Stop

What if the final terminus on many of these infiltration routes had been so close to an Allied base as to be literally in its shadow? Those with tunnels might have provided some interesting benefits — like total protection from shelling and uncontested entry to the bases. To keep out unwanted visitors, the areas around them would have been heavily mined. A few hundred meters to the north of the airstrip at An Hoa was an area so heavily booby trapped as to be routinely avoided by Marine sweeps and patrols.[13] Charles Soard, a former machinegunner with Golf Company, 2nd Battalion, 5th Marines, visited Duc Duc — the tiny village 200 meters from the northeastern end of the airstrip — in October of 1999. A VC veteran told him that to escape shelling she would move into an underground room and then up a tunnel that led beneath the Marines' barbed wire. She and her comrades had been hesitant to extend the tunnel for fear of being discovered.[14] Still once inside the three rows of protective barbed wire, a skilled infiltrator would have little trouble sneaking between bunkers and covertly accomplishing just about anything he wanted.

There is other evidence of these "ultimate" avenues of approach to Allied bases. Marble Mountain, atop which sat the largest Marine helicopter facility in Vietnam, was later discovered to be hollow and functioning as a VC staging area.[15] One morning in late 1966, a Marine lieutenant watched incredulously as a lone enemy sapper disappeared into a spider hole (or tunnel entrance) just outside the barbed wire along the Dong Ha airstrip.[16]

## The Approach March Formation

In Southeast Asia, enemy assault troops traveled in single file and widely dispersed. They opted for speed over flank security possibly because they were traveling over routes predetermined to be safe. Only the VC infrastructure could have provided such assurance.

During the actual movement, a reconnaissance-intelligence team preceded the formation at a distance of two hundred to three hundred meters depending on the terrain and weather. Local guides accompanied the recon element if it was unfamiliar with the area. This point element was followed by two combat units, the unit headquarters and heavy weapons section, combat support elements, and another combat unit. Following the formation at an interval of one hundred to two hundred meters was a rear-guard detachment from the trailing combat unit. This formation was consistent for squad through battalion with the only difference being the number of men in each element.

In territory where enemy contact was likely, individuals walked in file at intervals of five to ten meters during both day and night movements. . . . In larger formations, platoons traveled at fifty-meter intervals between each other with companies one hundred meters apart. . . .

. . . Although most moves were made at night, the soldiers had strict concealment procedures to follow in day moves. They camouflaged their uniforms and equipment with vegetation typical of the area and followed routes covered by overhead canopy.[17]

## The Conduct of the Actual Movement

The recon team and local guide normally spared the VC/NVA maneuver element from ambush or booby trap. Speed saved it from U.S. supporting arms. One evening in August of 1968 at Liberty Bridge (an outpost northeast of An Hoa), commanders of a Marine artillery battery and rifle company were scanning the countryside with long-range binoculars. At about 4000 meters to the east, they spotted 40 bushes running north along a paddy dike — obviously an enemy unit headed for Danang. After a single spotting round, they fired for effect into a clump of trees that had just swallowed the column. Without the normal delays of clearance and bracketing, they may have failed to destroy their target, but at least they came close.[18]

## Getting Resupplied Along the Way

Later than month and a mile to the south of the above sighting, three Marine battalions attempted to corner the 402nd VC

Sapper Battalion at the village of Chau Phong.[19] Near the southern half of this hamlet, a major trail exited the mountains. It was found to contain an overabundance of military materiel.

At 1500 [3:00 P.M. on 17 August], the 3rd Battalion, 5th Marines finally assaulted the hamlet [Chau Phong (2)], finding "many enemy dead, weapons, equipment, and food supplies." The enemy cache yielded significant quantities of

**Map 10.1: Danang's Southern Conduit**

(Source: 1:50,000 Map of Area Northeast of An Hoa Marine Base)

stores, including 88 tons of rice and enough medical supplies to support 500 men (5thMar CmdC, Aug68, p. 2-1).[20]

— *U.S. Marines in Vietnam: The Defining Year, 1968*
History and Museums Division, HQMC

The final engagement of this cordon operation occurred on 18 August some 1500 meters to the north of "Chau Phong (2)." In the contested village of "La Thap (1)," Golf Company, 2nd Battalion, 5th Marines found defender paraphernalia, rice bags camouflaged to look like the furrows of a field, and machineguns preserved in cosmoline. A few hundred yards to the south in what appeared to be a nonfortified area, the company found an underground room full of medical supplies — possibly a field surgery chamber.[21] It would appear, that the maneuver force spotted from Liberty Bridge early in August of 1968 had been moving along a route so heavily traveled as to have resupply points and evacuation facilities all along it. There is no telling what a detailed search of the microterrain along this conduit might have uncovered. But the big-picture thinkers had failed to grasp the significance of the little picture, so the cursory sweeps continued.

# 11 The Surprise Urban Assault

- How can Eastern forces so easily capture a city?
- Why cannot a Western-style defense stop them?

(Source: Courtesy of Cassell PLC, from *World Army Uniforms since 1939*, © 1975, 1980, 1981, 1983 by Blandford Press Ltd.; Corel Gallery, *Clipart* — Plants, 35A033)

## The Asian Realizes the Risk of Offensive Urban Combat

Sun Tzu disliked attacking cities. To do so, his descendants exercise great care.

> When surrounding cities, if the enemy side has few soldiers but plenty of food supply and external aid, one should conduct a quick attack.[1]
> — *100 Strategies of War*

First, the Asian commander occupies the city without appearing to do so. Then, he goes after the strategically important targets within that city — engaging anyone who gets in his way with ordinary forces and defeating them with extraordinary forces.

> Rule: Despite the enemy's high walls and deep moats, one should attack their strategic areas which they will have to defend.[2]
> — *100 Strategies of War*

Finally, by disrupting command/control and reinforcement/resupply, the Oriental commander tries to get the city's defenders to leave. He continues to attack only as long as long as his strategic gains outweigh his personnel losses. He bases his chances for success on two assumptions: (1) an unnoticed attacker will be tougher to stop, and (2) an already occupied objective will be easier to assault. One of the best examples of successful urban assault occurred at Hue City in February of 1968.

## The Flower Blooms

In Buddhist mythology, Hue was a lotus flower that had sprung from a mud puddle.[3] For a while the capital of Vietnam, it lay 100 kilometers south of the DMZ and at the narrowest part of Vietnam's coastal plain. Inside Hue was a massive fort called the Citadel. Square in shape and measuring 2500 meters on a side, its corners faced the cardinal directions. Its 6-square-kilometer interior was enclosed by walls 26 feet high and 40 feet thick. On three sides were canals; on the fourth, a river. Within the Citadel was another walled enclave — the Imperial Palace compound. It and the nearby Citadel flagpole may have been the focus of main effort for the 31 January attack. Both had symbolic significance to the city's defenders.

To take Hue City, the North Vietnamese may have used a Viet Minh technique known as the "blooming lotus."

> This methodology was developed in 1952 in an assault on [the town of] Phat Diem. Its key characteristic was to avoid enemy positions on the perimeter of the town. The main striking columns move directly against the center of the town seeking out command-and-control centers. Only

then were forces directed outward to systematically destroy the now leaderless units around the town. This outward movement, like a flower in bloom, gives the tactic its name.

This approach contrasts sharply with Western doctrine, which traditionally would isolate the town, gain a foothold, and systematically drive inward to clear the town. This sets up a series of attrition-based battles that historically make combat in built-up areas such a costly undertaking.[4]

Several assault columns may have moved simultaneously in the *Yingwei Battle Array* discussed in Chapter 4. Each column — or "urban thrust" — probably had as its objective an area containing several targets of strategic import. One enemy regiment (roughly one third the size of a U.S. regiment) had been assigned to the Citadel, and another to the more modern center of town.

At the same time that elements of the NVA 6th Regiment were moving into position to attack the Citadel, the NVA 4th Regiment came out of the mountains to the [south]west and began infiltrating the area south of the Perfume River. Darkness and heavy fog hid the movements of both enemy forces.[5]

To determine thrust destinations, one must only know which clusters of targets came under attack. Three clusters did inside the Citadel: (1) 1st ARVN Division Headquarters, Route 1 approaches, and northwestern-wall gates;[6] (2) Tay Loc Airfield, 1st ARVN Ordnance Company Armory, and southwestern-wall gates;[7] and (3) Imperial Palace, southeastern-wall gates, and northeastern-wall gates.[8] Three other clusters were hit outside the Citadel: (1) MACV headquarters, Provincial Treasury Building, Post Office, Public Health Building, and University;[9] (2) Provincial Prison, Municipal Power Plant, and Provincial Administration Complex;[10] and (3) Tu Do Stadium, An Cuu Bridge, and 7th ARVN Armored Cavalry compound.[11] A separate battalion had probably been tasked with each set. U.S. intelligence reports and subsequent events certainly support this hypothesis.

U.S. order of battle records listed the *6th NVA* [regiment] headquarters . . . in the jungle-canopied Base Area 114, about 20 to 25 kilometers west of Hue. . . . [T]he *4th*

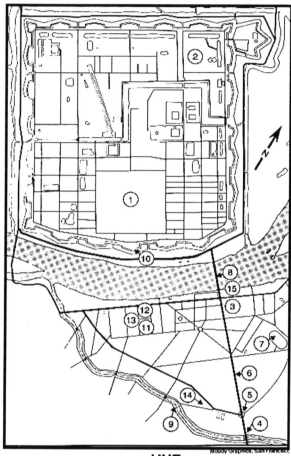

## HUE

1. Imperial Palace
2. 1st ARVN Division CP
3. MACV Compound
4. An Cuu Bridge
5. Traffic Circle
6. Canefield Causeway
7. Tu Do Stadium
8. Nguyen Hoang Bridge
9. Phu Cam Canal
10. Citadel Flagpole
11. Thua Thien Provincial Prison
12. Thua Thien Provincial Admin. Center
13. Hue Municipal Power Station
14. Hue Cathedral
15. Doc Lao Park

Scale 1 : 12.500

## Map 11.1: Hue City's Targets of Strategic Significance

(Source: Courtesy of Pacifica Military History, from *Fire in the Streets: The Battle for Hue, Tet 1968*, © 1991 by Eric Hammel)

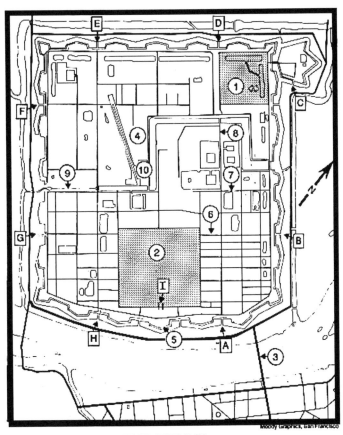

## THE CITADEL

| | | | |
|---|---|---|---|
| 1. | 1st ARVN Division CP Compound | A. | Thuong Tu Gate |
| 2. | Imperial Palace | B. | Dong Ba Gate |
| 3. | Nguyen Hoang Bridge | C. | Truong Dinh Gate |
| 4. | Tay Loc Airfield | D. | Hau Gate |
| 5. | Citadel Flagpole | E. | An Hoa Gate |
| 6. | Mai Thuc Loan Street | F. | Chanh Tay Gate |
| 7. | Tinh Tam Street | G. | Huu Gate |
| 8. | Dinh Bo Linh Street | H. | Nha Do Gate |
| 9. | Thuy Quan Canal | I. | Ngo Mon Gate |
| 0. | 1st ARVN Ordnance Company Armory | | |

## Map 11.2: Key Locations within the Citadel

(Source: Courtesy of Pacifica Military History, from *Fire in the Streets: The Battle for Hue, Tet 1968*, © 1991 by Eric Hammel)

*NVA Regiment* was in the Phu Loc area near Route 1 between Phu Bai and Danang (Enemy OOB, Anx. A, III MAF Periodic Intel. Rpt., No. 2, dtd 13Jan68).

Unknown to the Allies, both regiments were on the move towards Hue. The *6th NVA* had as its three primary objectives the *Mang Ca* [1st ARVN Division] headquarters compound, the Tay Loc airfield, and the Imperial Palace, all in the Citadel. South of the Perfume River, the *4th NVA* was to attack the modern city. Among its objective areas were the provincial capital building, the prison, and the MACV advisors' compound. The two regiments had nearly 200 specific targets in addition to the primary sites, including the radio station, police stations, houses of government officials, the recruiting office, and even the national Imperial Museum. . . .

By [the evening of 30 January] . . . the *6th NVA Regiment* was only a few kilometers from the western edge of the city. . . . At 2000 [8:00 P.M.], the regiment "resumed its approach march."

At this point the *6th NVA* divided into three columns, each with its particular objective in the Citadel. At 2200 [10:00 P.M.], about four kilometers . . . [from] Hue, the commander of the 1st ARVN Division Reconnaissance Company . . . and his men observed the equivalent of two enemy battalions filter past their positions, headed toward Hue. The two battalions were probably the *800th* and *802nd Battalions* of the *6th NVA* (Troung Sinh, "The Fight to Liberate Hue City," p. 90).[12]

— U.S. Marines in Vietnam: The Defining Year, 1968<br>History and Museums Division, HQMC

The North Vietnamese unit, augmented by area Vietcong *[sic]* troops, infiltrated through the Reconnaissance Company screen toward the Citadel's southwestern wall, where they waited for the signal to storm the city. When the signal came, they were aided by confederates inside the Citadel who dispatched any ARVN guards and opened the gates for them. . . .

Meanwhile the enemy positioned a third battalion, the 806th, outside the northwestern wall to form a blocking position astride Route 1 to prevent any reinforcements from reaching the city. The battalion's best company . . . had the mission of attacking the 1st Division Headquarters in the Citadel just over the wall. The invading force was bolstered by the NVA 12th Sapper Battalion.[13]

To enter the Citadel, one battalion of the 6th NVA used the Huu Gate, another the Chanh Tay Gate, and the other, the An Hoa Gate.[14] All relied on the same method to silence the guards.

After entering the Citadel through the sluice in the southwestern wall, four Communist soldiers — two members of a local VC unit and two NVA sappers — approached the Chanh Tay Gate. . . . The four, who were dressed in ARVN uniforms, were to overwhelm the gate guards and, from inside, open the way for the assault battalion lying in wait right outside the wall.[15]

There were two signals — one at 2:30 A.M. to take silent action, and the second at 3:30 A.M. to do what might make noise. Rockets and mortars will mask the sound of bangalore torpedoes and satchel charges detonating.

At 0233 [2:33 A.M.], a signal flare lit up the night sky above Hue. At the [south]western [Huu] Gate of the Citadel, a four-man North Vietnamese sapper team, dressed in South Vietnamese Army uniforms, killed the guards and opened the gate. Upon their flashlight signals, lead elements of the 6th NVA entered the old city. In similar scenes through the Citadel, the North Vietnamese regulars poured into the old imperial capital.[16]
      — *U.S. Marines in Vietnam: The Defining Year, 1968*
      History and Museums Division, HQMC

BAM! There was no mistaking what that [loud] noise was. . . . A 122mm rocket . . . had landed not more than 50 feet from where I was sleeping [in the MACV compound].
    . . . I . . . stumbled barefoot out the door toward the bunker. . . . There were about 12 of us in the bunker as two more booming explosions shook the compound. . . . [Then] I checked my watch. It was 0345 [3:45 A.M.].[17]
      — Capt. George W. Smith, U.S. advisor to ARVN unit

The 4th NVA Regiment came in from the southwest. The battalions tasked with MACV and 7th ARVN Armored Cavalry compounds got lost or delayed by Allied artillery south of the city.[18] The first Marines to reinforce MACV saw a large enemy column moving north parallel to Route 1 on the morning of 31 January.[19]

The third enemy regiment to enter the fight — the 5th NVA — came from the northwest.[20] Its job was to reinforce the Citadel.[21] Its forward base camp was probably in the La Chu Woods 5 miles northwest of Hue — near where U.S. Army units captured a three-story high and two-story deep bunker on 21 February.[22]

The VC column discovered by an ARVN sweep east of the city before dawn on 31 January was no ordinary patrol. The force was large, determined, and bent on controlling everything east of the Perfume River (its probable role in the overall battle will be discussed in Chapter 13). After being encircled, two ARVN battalions had to break out to the northeast and return by motorized junk to the Citadel. One arrived just outside the fort's northern corner at 3:00 P.M. on 1 February, and the other on 4 February.[23]

Some of the NVA battalions may have moved into the city not as whole units, but as separate companies or platoons that subsequently rendezvoused. As NVA maneuver elements routinely used local Viet Cong for reconnaissance and security, Hue's VC may have helped in a number of ways: (1) prestaging supplies, (2) guarding key intersections and bridges, (3) guiding columns into the city, and (4) opening the gates to the Citadel. While NVA sapper battalions often reinforced NVA infantry regiments, elements the VC Hue City Sapper Battalion were known to have helped the 4th NVA in their assault on southern Hue.[24]

## The Preliminaries

The North Vietnamese would not tackle a target like Hue without extensive preparation.

The Viet Cong [or NVA] fighting method was characterized as "one slow, four quick." The [one] slow was all the painstaking preparation that they put into any operation: repeated reconnaissance of the target, the building of a scale model of the objective so that the men assigned to the mission would recognize every feature, rehearsals of the planned attack as part of training, and the placing of arms caches and food dumps in forward areas. The four "quicks" followed when the operation actually began. First, there was the movement from the base area to the region of the objective, usually in small groups that would only reassemble just before they were to go into action. Then came the attack itself, where speed was the essence. The third "quick" was the clearance of vital arms from the battlefield and the retrieval of the dead and wounded. Finally, there was withdrawal. This was always scrupulously prepared as part of the original plan and depended heavily on a detailed knowledge of the local terrain and the position of enemy forces.[25]

## Reconnaissance

The NVA used two sources of information on when and where to strike. First, they encouraged the "revolutionary infrastructure" to provide information on the actions and locations of all defending forces.

Communist agents had actually infiltrated Hue 6 months before, organizing political cells and drawing up maps of Allied defenses.[26]

Then, the North Vietnamese sent reconnaissance teams into the city ahead of the strike forces. Their job was to update information and provide confirmation for the assault.

## Pre-Staging of Supplies

In all probability, 30 days of supplies had been smuggled into the city long before the battle. The Perfume River, Phu Cam Canal, and roads had been clogged with traffic throughout the month of January. This elevated level of activity provided ample opportunities for smuggling.

The weapons and ammunition were stashed throughout the city.[27]

Hundreds of infantry weapons — including .51 caliber heavy machineguns and perhaps hundreds of tons of ammunition, demolitions, and supplies — had been smuggled into Hue disguised as civilian goods.[28]

In addition to the tons of supplies that had been prestaged inside Hue, the participating regiments had resupply depots at their base camps just outside the city. One was located in the La Chu Woods northwest of Hue. Another (probably belonging to the 6th NVA Regiment) was discovered just west of Hue a week before the battle.

> The 2nd [ARVN] Airborne was lifted into an area about 8 kilometers west of Hue [around 22 January] to investigate reports of enemy activity.
> "We discovered a huge cave dug into the side of a hill. It was filled with all brand-new stuff, much of it still in boxes," Cobb [the U.S. advisor] said many years later. "We found dozens of machineguns, six 60mm mortars, 27 brand new SKS rifles with bayonets, surgical equipment, and 3 tons of rice. It was obviously a regimental headquarters."
> What they did not find were any enemy troops. "It's a good thing because we would have been badly outnumbered," Cobb said. "I think many of them were already in Hue disguised as civilians and reconnoitering the city."[29]

Of course, the enemy could also count on helping himself to the weapons and ammunition of his oversupplied opponent. He had at his disposal everything in the 1st ARVN Ordnance Company Armory and 7th ARVN Armored Cavalry tank park.

*Rehearsal*

Little is known about the rehearsals that preceded the attack on Hue except that C-ration boxes were turned into a miniature reproduction of the city. Practice sessions must have occurred at regimental base camps in the mountains. They probably involved a full-scale mock-up of a fortified urban compound.

What is known for sure is that every North Vietnamese participant of whatever rank knew what was being attempted and what his role would be — from strategic objectives to room clearing techniques.

## The Deception

The entire Tet offensive had, as its feint, the siege of Khe Sanh. The Allies did not believe that the enemy could mass 40,000 troops around Khe Sanh and still mount a countrywide offensive.

But the Asian is a master of ruses at every level. The extent of the deception at Hue was exceptional. At the time of the attack, there was a cease-fire in effect, ARVN forces on leave, strings of firecrackers going off, and jubilant automatic-weapons fire. The NVA even made sure Hue's defenders had become accustomed to the sights and sounds of Tet.

There was also a curious announcement from North Vietnam changing the start of the Tet holidays. Because of an unusual conjunction between the moon, the earth, and the sun, North Vietnam's leaders said the Tet holiday would not begin on 30 January, as indicated on the lunar calendar, but on 29 January.[30]

To make matters worse, heavy ground fog was expected throughout the Vietnamese coast plain during the predawn hours of 31 January 1968.

## Infiltration Technique

The Hue City Viet Cong had been helping to prestage supplies for months. Right before the attack, their job was to control bridges, road junctions, and gates to the Citadel. All nine gates fell almost immediately. Only the Hau Gate — the one right behind the 1st Division Headquarters compound — was expeditiously reopened.[31]

How many enemy soldiers were already inside the city when the assault came is not known. Each NVA battalion had probably already sent in its own reconnaissance personnel. Some of the battalions used VC companies to spearhead their attacks.

By the evening of January 30, VC spearhead companies had already slipped into the city in the guise of civilian pilgrims.[32]

The additional young men in Hue were thought to be ARVN soldiers on leave.

On 30 January, some of the enemy shock troops and sappers entered the city disguised as simple peasants. With their uniforms and weapons hidden in baggage, boxes and under their street clothes, the Viet Cong and NVA mingled with the Tet holiday crowds (Col. John F. Barr, Comments on draft, dtd 24Nov94). Many donned ARVN uniforms and then took up predesignated positions that night to await the attack signal (FMFPac, MarOpsV, Jan68, pp. 18-20, and Feb68, pp. 8-10).[33]

> — *U.S. Marines in Vietnam: The Defining Year, 1968*
> History and Museums Division, HQMC

Later, the U.S. advisor to an ARVN unit sent to recapture the city reported that many of the enemy never bothered to change back into their military uniforms.

"Many of the enemy we killed were wearing civilian clothes, which I guess made it easy for them to infiltrate the city,"

**Figure 11.1: Infiltration Opportunies in a Built-Up Area**
(Source: *FM 90-10-1* (1982), p. 2-3)

Chase said. "Another thing I noticed was the they didn't even bother to police up the weapons and ammunition from our troops. This made me believe they had everything they needed. They didn't need any of our weapons. They came to Hue well armed and with all the supplies they could carry."34

## Assault Technique

The NVA had its own methods for seizing a fortified urban area. When the company tasked with taking the 1st ARVN Division Headquarters compound discovered that an old water-gate bridge had been replaced with barbed wire, it resorted to a lengthy mortar barrage.35 To the hunkered-down defenders, B-40 rocket, bangalore, and grenade explosions all seemed like part of the indirect-fire attack. Then with a machinegun base of fire, soldiers armed with satchel charges tried to assault the main entrance.36 This was, in all probability, the routine feint by ordinary forces. While all of this was going on up front, General Truong saw enemy soldiers using grappling hooks to negotiate the rear walls.37

Although the enemy battalion penetrated the division compound, an ad hoc 200-man defensive force consisting of staff officers, clerks, and other headquarters personnel managed to stave off the enemy assaults.38

— *U.S. Marines in Vietnam: The Defining Year, 1968*
History and Museums Division, HQMC

(Source: Gallery Graphics, Mac/EPS, *Flowers, Trees, and Plants 2,* "Shrubs")

**161**

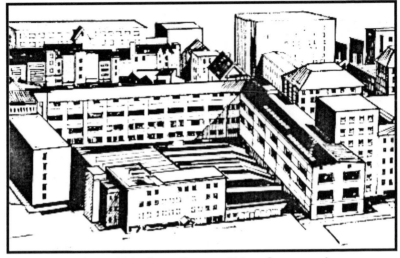

**Figure 11.2: Assault on an Urban Compound**
(Source: *FM 90-10-1* (1982), p. 2-3)

Back at the MACV compound, the attack sequence was about the same: (1) pinpoint-accurate 122mm rockets, (2) mortars, (3) B-40 rockets, (4) a frontal attack supported by a base of fire at the entrance to the compound, and then (5) a more secretive attack around back. Neither of the attempted penetrations succeeded, but something (possibly a satchel charge) scored a direct hit on a bunker out back.[39]

**The Consolidation**

As Allied forces began to arrive in Hue's southern outskirts, the 4th NVA regiment appears to have systematically withdrawn into the Citadel. It blew the An Cuu Bridge over the Phu Cam Canal three kilometers south of the walled city on 4 February and the Nguyen Hoang Bridge over the Perfume River on 7 February.[40] While moving into the Citadel on 16 February, one NVA battalion that had been fighting south of the Perfume River got hit by artillery.[41]

# 12   The Covert Urban Defense

- *How do Eastern defenders stay hidden in a city?*
- *Why do Western forces have trouble crossing streets?*

(Source: Courtesy of Cassell PLC, from *Uniforms of the Indo-China and Vietnam Wars,* © 1984 by Blandford Press Ltd.; Corel Gallery, *Clipart — Plants,* 35A017)

## Nobody Home

The literature talks very little about how to practice maneuver warfare in the city. One must draw some parallels to its rural application. The ways to defend might be about the same: (1) tank canalization, (2) standoff fire, (3) strongpoint matrix, and (4) repeated ambush. Their purpose would be to disrupt enemy momentum and facilitate one's own counterattack. Concurrent with the physical aspects of the battle would be an effort to get into the attacker's mind — to confuse and demoralize him. This might be accomplished by stubbornly defending and then abandoning successive pieces of nonstrategic real estate. In urban terrain, the profusion of protected passageways makes systematic withdrawal much easier.

Like the multi-tiered preparation of Iwo Jima, the maneuver war defense of a city might include the following: (1) unoccupied ground covered by long-range fire; (2) belts of well-hidden, mutually supporting bastions; and (3) defiladed conduits of reinforcement/withdrawal. Forward positions would be carefully camouflaged, covered by fire from behind, and conducive to rapid abandonment. In a built-up area, the open ground would be vacant lots and streets, and the terrain features would be whole blocks and durable buildings.

Some of the hardest street fighting in U.S. history occurred at Hue City in February of 1968. For interesting reasons, the North Vietnamese were able to defend its ancient Citadel for weeks.

## Limiting U.S. Resupply and Reinforcement

Characteristic of any type of maneuver warfare operation, whether offensive or defensive, is isolating the battlefield — preventing the enemy from being resupplied or reinforced. By maintaining relative parity of wherewithal, the side with more combat skill improves its chances of winning. Over the years, adversaries have frequently interrupted the flow of U.S. materiel to better utilize their strong suit — small-unit tactics.

On the fourth day of the battle for Hue City, "North Vietnamese sappers blew the An Cuu bridge, closing the land route into the city."[1] All Allied supplies and reinforcement had to come into the Citadel by boat or helicopter through a holdout ARVN compound at its northern corner. Those boats and helicopters came under fire. When that fire found its mark, Allied offensives stalled.[2]

A price was payed [sic] for getting the needed supplies to Hue. Sixty helicopters were hit or shot down over the city; many came back to base with dead or wounded crewmen. The worst hit were the Navy boats plying the Perfume [River]. NVA attacks in the form of rockets, mines, and ambushes to and from Hue, reached such intensity that, at the beginning of the third week of the battle, General Abrams requested [for river traffic protection] . . . patrol

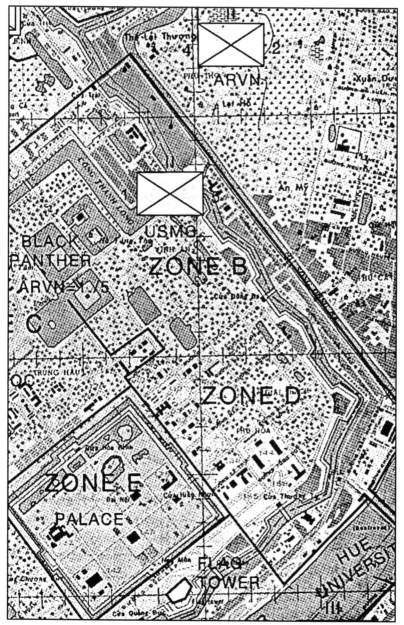

**Map 12.1: Eastern Corner of the Citadel**

(Source: *U.S. Marines in Vietnam: The Defining Year 1968*, Hist. & Museums Div., HQMC, p. 165)

boats . . . , helicopter gunships, aircraft, artillery, and ground security troops. But the NVA ambushes were never entirely thwarted; bringing supplies to the Marines in Hue, one ammunition-filled LCU [Landing Craft Utility], and two LCUs carrying fuel bladders, were blown up [heading for the Citadel].[3]

When the 1st Battalion, 5th Marines (1/5) arrived by LCU at the river dock just north of the Citadel, they came close to experiencing the NVA's favorite way of isolating a battlefield.

Although encountering an occasional RPG round or enemy sniper fire from both banks of the Perfume River while on board the Navy craft, the Marines landed at the ferry landing north of the city without incident. As the troops were about to start their march to the Citadel, Major Thompson [the battalion CO] later related that "villagers warned me that the NVA had set up an ambush along the route I had chosen." The Vietnamese civilians guided the Marines along another road.[4]

## Carefully Choosing Where to Fight

A maneuver warfighter thinks hard about where to confront a more powerful attacker. He must hide until the last moment and then stop that attacker decisively. To do so in the city, he requires the distant views and gunfire protection of tall durable structures. From their upper stories, his artillery/mortar observers and machinegunners can control the surrounding streets for hundreds of yards. Some of those structures — like the steel-framed, corrugated-iron-covered grain elevator at Stalingrad[5] — are almost impervious to bombardment. Others are so massive as to permit internal tunneling and bunkering.

To the North Vietnamese, Hue's ancient inner city provided what would soon prove to be ideal defensive terrain.

[T]he [Allied] units inside the Citadel were confined in their movement by a number of structures and features that could not be crossed or flanked. The largest and most troubling of these was the Citadel wall. . . .

Another factor limiting Vietnamese and American units was the Imperial Palace. . . . For reasons of culture and politics, the palace was considered sacrosanct by the GVN [Government of Vietnam] forces and their allies. NVA and VC firing down from the palace walls could be engaged only by small arms. Moreover the palace sat roughly in the center of the southeast half of the Citadel, channeling the attacking units into two distinct corridors that could not be mutually supported. In effect, in the southeastern half of the Citadel there were two distinct battlefields, isolated from one another and dominated by high walls on either flank.[6]

## Temporary Strongpoints

To defend, the maneuver warfighter often depends upon staggered rows of semi-independent "strongpoints." From these bastions, he can control unoccupied areas with long-range fire. In essence, the areas between the bastions become "firesacks." While assaulting what were thought to be enemy "lines" over the years, U.S. infantrymen have more than once encountered dummy positions and then automatic-weapons fire from both flanks.

Hue City's defense consisted of rows of hastily prepared strongpoints — structures and blocks from which the streets could be controlled by fire. As these strongpoints were arrayed in depth, any one of them could be abandoned without immediately jeopardizing any other. Every time a defensive position was penetrated, its occupants would move backwards to their alternate.

As they had during 2/5's final sweep between the Perfume River and the Phu Cam Canal, the NVA [inside the Citadel] were resorting to delaying tactics consisting of the unyielding defense of strongly fortified, mutually supporting strongpoints.[7]

Inside the Citadel, these bastions were particularly well designed. They contained interconnected, mutually supporting bunkers.[8]

The area held by the NVA was so compact that 1/5's attack

was less the reduction of a built-up city than an assault on a fortified position. Major Bob Thompson said later that the fighting during this period was more like that in Tarawa than that in Seoul. The whole objective was really one big strongpoint, complete with interlocking bands of fire from mutually supporting fortified positions.[9]

So ingenious were the defenses along one street — Mai Thuc Loan or "Phase Line Green" — that an air-, naval-gunfire-, artillery-, and tank-supported U.S. battalion took over four days (from early in the morning of 13 February to late in the afternoon of 16 February) to cross it.[10] How the enemy accomplished this amazing feat deserves closer inspection.

## Taking Out the Opponent's CP Group

Proponents of this alternative style of warfare have as one of their highest priorities disrupting the command-and-control apparatus of their adversaries. In other words, they will go after stationary command bunkers on offense and mobile command posts on defense.

Near Phase Line Green inside Hue City's ancient fort, U.S. Marine commanders soon learned how not to appear too obviously in charge.

As Captain Jim Bowe searched for a solution [to the fire from the Dong Ba tower] and waited for Battalion to decide what to do next, an enemy RPG team *sneaked* [italics added] into position opposite the Alpha/1/5 CP group and let fly one B-40 rocket. Captain Bowe, the company exec, the company gunny, and just about everyone around them was [sic] injured in the blast.[11]

Hunter-killer teams may have helped to control the unoccupied areas between strongpoints in Hue City, just as semi-independent squads had successfully defended the enclaves along the Volga at Stalingrad.[12] By moving from window to window, the NVA soldiers must have appeared more numerous than they really were. The Marines never guessed that some blocks were underdefended.

## Armor Poses Little Threat

In the dangerous canyons of a city, direct-fire weapons normally engage their targets from a distance. The maneuver war defender has several ways in which to discourage standoff fire: (1) occupying structures masked by other structures, (2) ambushing opposition armor, and (3) channeling enemy tanks into preplanned kill zones. In an Asian city, Western tankers don't get a clear shot at an enemy stronghold for long. They must concurrently endure dozens of antitank rockets from different directions. Seldom can they determine the location of any one of the shooters. Each antitank man must randomly fire and then rehide.

Inside the Citadel, roadblocks were encountered just as they had been in Seoul.[13] Canalized by the piles of debris and thick walls, the tanks and 106mm-recoilless-rifle-carrying ONTOs were forced to operate in less than favorable conditions. They could not fire at anything from long range because of the buildings in the way. Nor could they approach Mai Thuc Loan without coming under a multidirectional hail of RPGs or B-40 rockets. They had to charge up to Phase Line Green, take a few hastily aimed shots at the Dong Ba Tower or some building, and then quickly withdraw for repairs.[14] These hard-earned shots had little effect on the opposition. The hail of B-40s had done their job.

> [T]he tanks had problems getting around in the narrow
> streets. Crewmen were killed and wounded. Every tank
> had at least a dozen B-40s slam into it during the course of
> the battle. But, every night, they would pull back to the
> ARVN HQ, repair, and be back on the streets in the morn-
> ing.[15]

## What Only Appears to be a Cross-Street Ambush

In urban terrain, a maneuver warfare defender will remain so well hidden on one side of a street that an unwary opponent may fully expose himself on the other. Only then will his final protective fires look to that opponent like an ambush. Eastern adversaries have long known that U.S. soldiers are trained to immediately assault out of a close-proximity ambush.

Well that's precisely what happened on Mai Thuc Loan street on the afternoon of 13 February 1968,[16] except that the Marines had enough sense to move backwards instead of forwards. Phase Line Green had the Dong Ba Gate/Tower at one end and the Imperial Palace at other. The deadly combination of machinegun crossfire from the flanks and rocket and small-arms fire from the front must have been enough to trigger the Americans' common sense.

> When Charlie/1/5 jumped off at 1255 [12:55 P.M.], it ran straight into NVA automatic-weapons fire bolstered by . . . Chicom grenades and many B-40 rockets. The NVA had . . . entrenched themselves within and around many of the buildings facing the Marine company, and they were able to fire their weapons from virtually any angle into all parts of the fragmented platoon formations, including the rear.
> . . . Devastating .51-caliber machinegun crossfire was the most dangerous hurdle in the crowded streets, where reverberating echoes made it almost impossible for the Marines to locate the source of the enemy fire, ricochets were as effective as direct hits, and flying masonry chips were as injurious as bullets.[17]

By firing RPGs and B-40 rockets through window openings, the NVA could make the front rooms of the buildings across from them virtually untenable. The projectiles would explode against the back walls, wounding anyone in the room.[18]

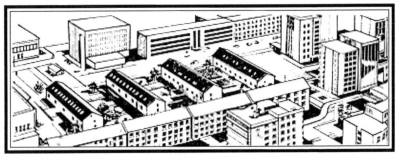

**Figure 12.1: Street Swept by Distant Machinegun Fire**
(Source: *FM 90-10-1* (1982), p. 2-4.)

## Streets Defended from Afar

A city street provides an extended field of fire. It cannot be crossed until the machineguns at either end have been neutralized.

While attacking toward the southeastern end of Hue's ancient fort, the Marines received continual fire from the high ground on both flanks — the Citadel and Imperial Palace walls. "Thus, as long as NVA machineguns and snipers could fire into 1/5's rear and flanks with virtual impunity, 1/5 could not advance."[19] Of course, the key to the defense of Mai Thuc Loan was the tower above the Dong Ba gate. During the afternoon of the first day, C Company tried three times to cross this street and only managed to get two people stranded in the final attempt.[20] The next day, the other companies gave it a try.

> Delta Company had made no progress whatsoever in trying to take over the dominating tower at the east end of phase line green [14 February]. The NVA occupying the tower had terrific fields of fire down the street . . . [and] their crossfire was devastating to the Marines who charged across the street. . . . Twice Alpha charged, and twice they were turned back, getting absolutely nowhere, at a terrible loss of life.[21]

## A Totally Comprehensive Fire Plan

Ammunition-rich Americans have designed fire plans in war, but nothing like those of their poorly supplied foes. To win before their ammunition runs out, those foes must cover every inch of ground by fire.

What the Marines faced on Mai Thuc Loan almost defies description. The U.S. had once again underestimated its opponent.

> They employed every trick in the book. For example, the NVA placed machineguns in sandbag or rubble bunkers built against the rear walls of masonry houses. They fired the guns through open front doorways, the better to obscure muzzle flashes and defy observation. They fired B-40 rockets, AK-47s, and SKSs from similarly covered posi-

171

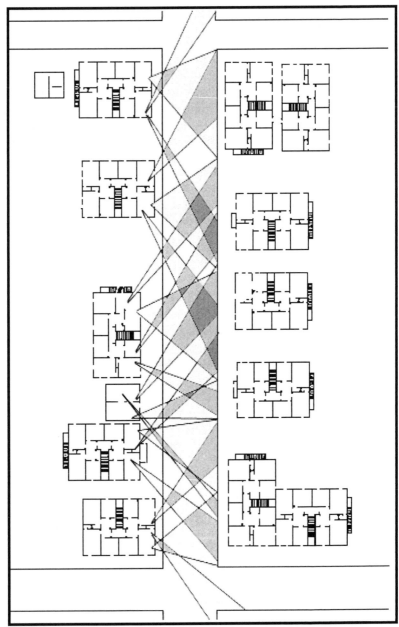

**Figure 12.2: Thin Bands of Recessed Fire**

(Source: *TC 90-1* (1986), p. 3-53)

tions. The only way to get a direct hit on an in-house bunker was to stand in the front doorway, exposed to fire from the target and from numerous spider holes covering every intersection and every opening between the houses. . . . In his years at the [Basic] school, [Major Bill] Eshelman had never heard of an integrated defensive plan as comprehensive as what he saw on the ground in the Citadel. He was frankly awed by the professionalism of his adversaries, who, he now realized, he had seriously underestimated.22

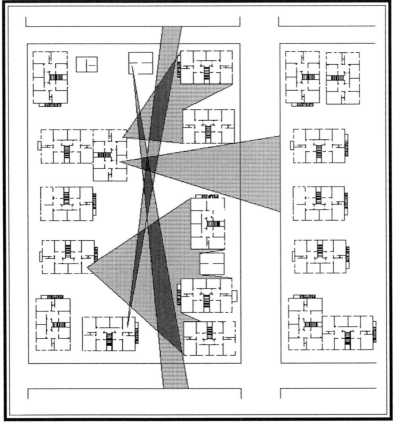

**Figure 12.3: Withdrawal Covered by Fire**
(Source: *TC 90-1* (1986), p. 3-53)

## What Appears to be a Rigid Defense Really Isn't

The maneuver warfighter seldom retreats but routinely withdraws to acquire a tactical advantage. Urban terrain offers safer avenues of egress than its rural equivalent. Backward movement can easily be accomplished through openings between buildings, interconnected attics or rooftops, sewers, tunnels, trenches, or any combination thereof.

If the Japanese had upgraded the Citadel's defenses during WWII, there is no telling how many below-ground conduits between buildings may have existed before the NVA even showed up. One thing is certain, the visitors from the North improved upon what had already been formidable fortifications.[23]

On the narrow streets inside the Citadel, the houses and stores were close together. For covered withdrawal, the NVA not only had the narrow spaces between buildings, but also trenches.[24] And when it was time for the occupants of a hard-pressed NVA position to pull back, they could do so under the protection of overhead fire. When one Marine squad finally crossed Mai Thuc Loan during A Company's third attempt on 14 February, it almost immediately encountered fire from behind the house it had entered. This fire had to be coming from an NVA reserve force at the back of the block. It had been tasked with limiting penetrations and covering withdrawal.

> [The] Alpha [squad] had rushed across the street en masse into the corner house. . . . Unfortunately, the squad had immediately come under fire as the NVA on the second floor of the house shot at them down the stairway. . . . At the same time, the Alpha squad leader was hit by a machinegun that raked the first floor from the backyard of the house next door, and two more Marines were hit by fire from yet another machinegun shooting at them from directly behind the house.[25]

Then a 1/5 mortarman, of all people, came from several blocks back to the aid of the beleaguered squad and single-handedly knocked out both enemy machineguns. At that point, the rest of the platoon rushed across the street to consolidate the beachhead.[26] But it was too dangerous for any more Marines to cross Phase Line Green until the Dong Ba Tower was taken.

## U.S. Bombardment Does Little Good

Preparatory fire does even less damage in the city than in the country. Walls and roofs will effectively block shrapnel.

At Hue City, the walls and towers of the Citadel were large and durable. Their defenders had tunnels through which to escape U.S. bombing and shelling.

> The Citadel wall . . . [was] as many as seventy-five meters thick in places and honeycombed *throughout* [italics added] by passages and bunkers excavated by the Japanese near the end of World War II.[27]
>
> The NVA had dug in at the base of the wall there and "tunneled back underneath this structure."[28]
> — *U.S. Marines in Vietnam: The Defining Year, 1968*
> History and Museums Division, HQMC

After taking a beating for a day and a half on Phase Line Green, all but the A Company Marines pulled back to unleash artillery and air strikes on the tower.

> The only reasonable alternative to a toe-to-toe struggle between infantry in the streets was a voluntary withdrawal by 1/5 so massive supporting arms could be employed. Major Bob Thompson and Major Len Wunderlich drew up a plan, got approval for it from Regiment, and began easing back from the fight along Mai Thuc Loan. The planned artillery and naval gunfire bombardment could not be initiated until all friendly troops were clear of the impact zone, and that did not happen until the middle of the afternoon [14 February].[29]

Unfortunately, much of what the Marines had been told about the effect of supporting arms on a poorly equipped opponent would soon prove false.

> And now the F-4s were coming to Hue, finally, to try to blow the NVA out of the [Dong Ba] tower. Sitting on that wall, waiting for the show, I knew with an absolute cer-

tainty that the NVA in the tower were doomed, and that we would have control of this critical piece of high ground very soon, or that it would no longer be high ground.

I was wrong on both counts.

Oh, the F-4s came, and the F-4s dropped stick after stick of napalm, 250-pound snake-eye bombs, and 500-pound high-drag, high-explosive bombs on the tower. And nearly every bomb dropped was a perfectly placed direct hit. The awesome firestorms from the detonating napalm canisters engulfed the tower, burning every square inch of the tower's surface; the high-explosive bombs pounded the tower and eventually reduced its height by at least ten feet. But the NVA did not run away; *few of them* [italics added] died; and every single bombing run made in the F-4 pilots' valiant attempts to destroy our enemy entrenched in the tower was greeted by a high volume of small-arms fire from the enemy.

The NVA AK-47s and .30-caliber machineguns ripped upward, directing their light green and white tracer rounds at the invading Phantoms. The enemy's constant and defiant small-arms fire was only momentarily interrupted exactly when the napalm burst into a whooshing roar of flame and smoke and exactly when the high explosives burst their gut-wrenching concussive power on the tower. Immediately after, the NVA gunners stuck their . . . heads back up and started shooting again at the flame-spewing dual exhausts of the departing Phantom's jets. The Phantoms made pass after disciplined pass, dropping no more than two bombs at a time, but the determined NVA gunners survived all of them and always had the last word in the deadly duel. One Phantom even took a couple of hits up one of his tailpipes and had to limp back to Danang without dropping all his ordnance on the tower.[30]

That the wall and towers contained a virtual labyrinth of subterranean bunkers and tunnels should by now be apparent. The Dong Ba Tower had, after all, been built over a gate. Even during successive ground assaults, "the best NVA positions were remanned as quickly as they were knocked out."[31] What the Japanese had started, the NVA had obviously improved upon and then put to good use. One wonders how many maneuver warfare techniques may have circulated around the Orient but never reached the Western World.

## The Embrace

As U.S. forces have pulled back over the years to safely unleash their seemingly unlimited arsenal of supporting arms, foes have often followed closely behind them to escape destruction.

When the Marines pulled back from Phase Line Green to use airstrikes and artillery, their opponents moved forward. When the Marines tried to return to their original position, they ran into unexpected resistance.

According to Thompson, the NVA also moved forward when the Marines fell back to use their supporting arms, "so when the fires were lifted we had to fight to retake more ground."[32]

Bravo/1/5 and Charlie/1/5 were able to fight their way back to Mai Thuc Loan Street [on February 15] primarily because the NVA conceded the ground. Delta/1/5 was fought to a standstill well short of Mai Thuc Loan and the Dong Ba tower.[33]

## Spoiling Attacks from the Rear

Opponents have been quick to realize the power of spoiling attacks from the rear. The relative ease of infiltration through urban terrain creates plenty of opportunity.

Inside the Citadel at Hue City, the more obvious spoiling attacks took the form of mortar or sniper fire. Luckily for the Marines, the enemy never mounted a concerted night infiltration attack around the exposed right flank of 1/5.

Father McGonigal's body [1/5's adopted chaplain] had been found in the rubble of a house two blocks behind the front line [18 February]. No doubt, while on one of his countless errands of mercy, the priest had been caught in one of the mortar barrages the NVA fired into 1/5's rear every evening.[34]

Several sniping incidents behind 1/5 convinced Major Thompson that there were NVA and VC infiltrators among the refugees, but there was nothing he could do about it.[35]

## Tactical Withdrawal to Prepared Fallback Positions

Unlike the U.S. soldier, a maneuver warfare defender depends for his ultimate survival on a timely withdrawal.

The enemy in Vietnam could not be dehumanized as easily as he had been in the Pacific. In the late sixties, people who routinely rushed machineguns and only occasionally took prisoners had more difficulty claiming that their opponents had been "fanatical." While there were rumors early in the war of enemy soldiers chained to their holes and high on drugs, there is little proof of it. When any building along Phase Line Green became untenable, the NVA simply pulled back. But they did not abandon Mai Thuc Loan until the Dong Ba Tower could no longer be successfully counterattacked.

It turned out that Delta Company would need two more days, several attacks, and the heroics of . . . Delta Company Marines to finally take the tower the first time. Shortly after that, the NVA counterattacked, and Delta had to withdraw in a hurry, dragging their wounded back with them. Delta would attack and take the tower four separate times, only to have the NVA take it back again. Finally, on the fifth day of fighting, Delta would finally seize and hold the tower, after having taken terrible losses. Obviously, the NVA understood that this tower was the high ground in this battle and that whoever held the tower could easily keep their enemy pinned down. The NVA were not going to give up the tower just because of the annoying Phantoms.[36]

Early in the morning of 16 February, Bravo Company was taken out of reserve status and committed to the battle. They were ordered to move across phase line green to assist Alpha Company in securing Alpha's control of the block, and by the end of that day, the combined force of two companies of Marines had forced the NVA to escape south across phase line orange, or Nguyen Bieu, the next parallel street south of phase line green.[37]

The frontline NVA defenders didn't pull back without a plan. They withdrew methodically (under covering fire from reserve units behind them) into prepared positions two blocks back, just as in the two-line defense system discussed in Chapter 8.

> With little room to outflank the enemy [in fighting outside the Citadel], the battalion [2/5] had to take each building and each block one at a time. According to [battalion commander] Cheatham, "We had to pick a point and attempt to break that one strong point." After a time, Cheatham and his officers noted that the enemy "defended on every other street." In other words, the battalion would move quickly and then hit a defensive position.[38]

The second cross-street after Phase Line Green — "Phase Line Black" — put up more resistance than the first — "Phase Line Orange."[39] But without covering fire from the Citadel wall, the NVA could not hold these streets for long.

### Eluding Return Fire

Eastern soldiers have tried harder to stay hidden than their Western counterparts. Why is this? Both are equally smart and interested in longevity. The answer must lie in how much initiative they have been allowed.

The Oriental flair for deception was at work at every level in Hue City. Opposition positions were as difficult to locate as on Iwo Jima and for similar reasons. Knowing that human beings expect trouble from the same elevation and terrain, the NVA had shifted above or below ground and away from buildings. Here's what the Marines of C/1/5 found when they finally got across Mai Thuc Loan:

> Most of these houses were one-story homes, but a couple were two-story affairs, providing excellent and advantageous firing positions for the waiting NVA. From these

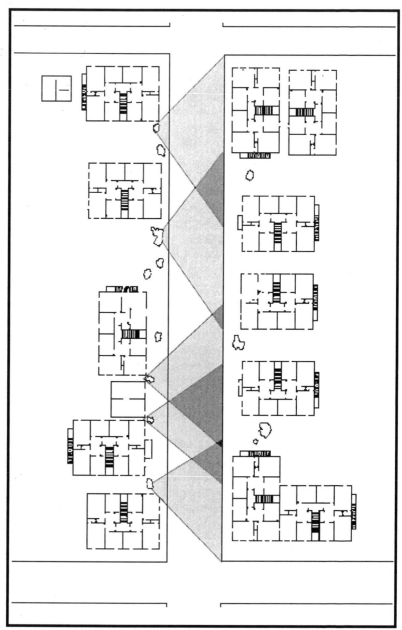

**Figure 12.4: Fire from Bushes between Buildings**
(Source: *TC 90-1* (1986), p. 3-53)

positions, the NVA could shoot right down on us, point-blank, as we tried to run across the street. This was obvious, and we understood the situation clearly, so we had directed our return fire at the windows and doorways of the houses across the street, which were the likely enemy firing positions. What we had not realized was that the NVA were also shooting at us from well-concealed, dug-in positions *between* the houses, at street level. . . .

. . . The near-solid frontage of homes that lined the south side of Mai Thuc Loan provided excellent defensive positions for the NVA defenders. The layout for the street also provided the opportunity to prepare well-camouflaged, almost invisible positions between the houses at the base of the thick, ancient foliage growing between the houses. For four days, we had been directing our fires at the windows and doors, and although we had most certainly been taking enemy fire from those positions, we had also been taking deadly enemy fire from ground level, from the fighting holes dug in under the bushes between many houses.

The NVA had cleverly and energetically prepared these positions well in advance of our arrival, and from the looks of those that I inspected, the NVA gunners who had occupied them had scored very well with nearly total impunity. There were no evident blood trails in the hidden foxholes that I inspected that morning, as there were in many houses we had passed through.[40]

Many of the Marines had been shot from above. Some of the NVA had moved from place to place on the rooftops.[41] But much of the deadly fire may have come from snipers in attics. It is not difficult, after all, to fire from the deep shadows of an attic, through a small opening, and into what is clearly illuminated outside. Then there would be no evidence of where the shot had come from or any way to hurt its perpetrator. Some of those snipers may have hidden behind the Marines and only fired when other shooting masked the noise.

[After crossing Phase Line Green] Charlie Company spent the rest of that day making sure that the blocks we had finally gained were completely secure. Every house was checked and rechecked, including the dusty, dirty attics, because we were fearful that an attempt might be made to hide a few NVA snipers, who could cause terrible chaos and confusion if they got behind our lines, especially at night.[42]

## Once Again

Detailed discussions of how the Japanese defended Manila, or the North Koreans defended Seoul, are difficult to find. Of interest, the North Vietnamese defense of Hue City closely resembles the "purported" last stand on Iwo Jima. When the Marines entered the Gorge at the north end of the island near Kuribayashi's cave, they encountered multi-tiered terrain very similar to that of a city.

In attacking these positions [in the Gorge], no Japanese were to be seen, all being in caves or crevices in the rocks and so disposed as to give an all-around interlocking, ghost-like defense to each small compartment. Attacking troops were subjected to fire from flanks and rear more than from their front. It was always difficult and often impossible to locate exactly where defensive fires originated. The field of fire of the individual Japanese defender in his cave position was often limited to an arc of 10° or less; conversely he was protected from fire except that coming back on this arc. The Japanese smokeless, flashless powder for small arms, always advantageous, was of particular usefulness here. When the position was overrun or threatened, the enemy retreated further into his caves where he usually was safe from gunfire, only to pop out again as soon as the occasion warranted unless the cave was immediately blown.[43]

— 5th Mar.Div. Intelligence Report of 24 March 1945

## Demoralizing the Attacker

The whole goal of maneuver warfare is to demoralize one's opponent. This can be accomplished in any number of ways. Among the most popular are to confuse one's adversary, injure him, and then create the impression that his return blow has had little effect.

At Hue City, the North Vietnamese worked on American minds by creating casualties while remaining invisible themselves. The Marines got little opportunity to witness any counterdamage (if, in fact, there was much done).

> Up until this moment [while looking through a Marine sniper's peephole after crossing Mai Thuc Loan], the NVA had represented a sort of mythical enemy, like their Viet Cong brethren. They had moved fast, hit hard, and eluded us seemingly at will. We had taken terrible casualties that first day on phase line green. We had taken many casualties as Marines exposed themselves to help their dead and badly wounded buddies lying exposed in the streets. I suppose we had killed or wounded a few NVA with our counterfire, and certainly many of them in the tower had been hit. But with the exception of the now squashed NVA body in the street, I had yet to see any dead NVA bodies.[44]

(Source: *FM 21-76* (1957), p. 56)

## Hidden Channels of Movement

Throughout the battle for the Citadel, the NVA may have maintained several entrance and exit corridors. As the mountains lay only four kilometers away, the initial conduits for men and supplies probably lay to the west.

While the battle for the Citadel raged, U.S. 1st Cavalry and 101st Airborne brigades worked hard to seal off the western approaches to Hue City.[45] Unfortunately, the NVA seldom departed the same way they came. Because there may have been more than one way into the city, U.S. efforts to isolate the Citadel may have met with limited success. And where way stations conceal subterranean facilities, what goes on above ground is only part of the story.

# The Vanishing Besieged Unit

● *How can a totally encircled Eastern unit just disappear?*

● *What are the details of its escape?*

## The Final Insult

Over the years, U.S. troops have entered many an enemy stronghold only to find it virtually unmanned. Helping to explain this incongruity throughout World War II and Korea were reports that the opposition had been pulverized by bombardment or trapped below ground.

Then in Vietnam, this unfortunate paradox was further dispelled by rumors that the enemy had been dragging off his dead. Still, enlisted veterans of that war remember all too well the shortage of enemy bodies after fully encircled pockets of resistance had been finally reduced. Upon reaching the southeastern wall of the Citadel on 23 February 1968, the 1/5 Marines could only deduce that their adversaries had withdrawn into the Imperial Palace.[1] But when the 1st ARVN Division's elite Black Panther Company stormed that Palace on 24 February, they found it empty.[2] Once again, an opposition unit — though totally surrounded and seemingly out of options — had mysteriously disappeared. Unique to this case is only the fact that the enemy unit had been a division.

## What Had Happened Soon Became Apparent

The fourth "quick" in the NVA attack method was, after all, withdrawal. As early as 16 February, intercepted radio traffic revealed the enemy commander's desire to withdraw his troops from the Citadel.[3] That request may have been only temporarily denied. For the NVA, final withdrawal was preceded by a holding action.

Despite the orders to stay and fight, it was becoming clear that even though the NVA forces were not quite ready to leave the Citadel, they were adjusting their strategy from a defense in place to a rear-guard action.[4]

The empty palace on 24 February should have come as no surprise. Citadel vintage forts often had built-in, underground passageways between their royal compounds and moat.[5] But still one wonders how such a large enemy force had managed to move unnoticed through the Allied cordon. A detailed examination of the U.S. battle chronicles for 22-25 February 1968 may provide valuable clues.

To the west of the American Marines, however the North Vietnamese continued to fight for nearly every inch of the old city still in their hands. In the Vietnamese Marine sector on the 22nd, the enemy fired 122mm rockets followed by ground attacks on the Marine positions. . . .
. . . ARVN and American artillery, on the night of the 23rd spoiled another NVA attempt to break through South Vietnamese defenses in the western sector of the Citadel. The 2nd Battalion, 3rd ARVN Regiment then launched its

own surprise attack along the south[east]ern wall. At 0500 on the 24th, soldiers of the ARVN battalion pulled down the Viet Cong banner . . . on the Citadel flag tower [near the wall]. By 10:25 P.M. on the 24th, the 3rd ARVN Regiment had reached the south[east]ern wall and secured it. Meeting little resistance, the ARVN troops, by late afternoon, recaptured the palace with its surrounding grounds and walls. . . . By nightfall, only the southwest [southern] corner of the Citadel remained under enemy control. Under cover of darkness at 0300 on the *25th* [italics added], the Vietnamese Marine Battalion launched an attack and eliminated this last pocket of North Vietnamese resistance in the Citadel. Outside the [north]eastern walls of the Citadel, a two-battalion ARVN Ranger task force cleared the Gia Hoi sector, a small enclave that had been under NVA control since 31 January. Save for mopping up operations the fight for the Citadel was over (Maj. Talman C. Budd II, MAU, NAG, CAAR, Hue City, dtd 25Jul68).[6]

> — *U.S. Marines in Vietnam: The Defining Year, 1968*
> History and Museums Division, HQMC

Of particular interest here is that, on the nights of 21, 22, and 23 February, the enemy still controlled the area between the eastern side of the Citadel and the Perfume River.

The Allies were not oblivious to the possibility of a breakout, but most of their units had been deployed to the west and north of Hue — the directions from which the NVA had come. That's how the 1st Cavalry Brigade got embroiled in the fight at La Chu Woods. But the NVA routinely withdrew along routes different from those used for approach. They liked to do the opposite of what was expected. Perhaps, they had planned all along to evacuate casualties and withdraw along routes never before used.

### The Southern Corridor

On 26 February, 2nd Battalion, 5th Marines had to fight hard for portions of a 2000-yard ridgeline running due south from the confluence of the Phu Cam Canal and Perfume River — from the southern corner of the Citadel.

**Map 13.1:  The Corridor South of Hue**

(Source: *U.S. Marines in Vietnam: Fighting the North Vietnamese in 1967*, Hist. & Museums Div., HQMC, p. 212)

On the morning of the 26th, the Marine battalion continued to attack to clear the ridgeline. . . . About 500 meters to the north, Company H . . . maneuvered to take the last hill on the ridgeline, where the enemy remained entrenched in fixed positions. About 1330 [1:30 P.M.], enemy defenders . . . forced Company H to pull back and call for an air strike. . . .

Resuming the attack after the air strike, Company H once more pushed forward. Again the Communist troops resisted the Marine advance. About 1620 [4:20 P.M.], once more unable to make any further headway, the Marine company called upon air to take out the enemy defenses. . . .

On the morning of the 27th, Marine air and artillery bombarded the enemy defenses. After the last fires had lifted, all three companies of the 2nd Battalion rushed forward. Reaching the crest of the hill without encountering opposition, the Marines discovered that the enemy had departed during the night. Strewn around the hillscape were 14 enemy bodies.[7]

— *U.S. Marines in Vietnam: The Defining Year, 1968*
History and Museums Division, HQMC

"Ridgelines" are a rarity in the lowlands of coastal Vietnam. One radiating straight out from an imperial palace suggests "escape tunnel." But whether or not this particular ridgeline contained a tunnel is immaterial. Its length remained open to enemy travel until 26 February.

During the fight for this ridgeline, the 2/5 Marines took fire from a Buddhist pagoda to their south[8] — quite possibly the famous Temple of Heaven ("Obj. 3" on Map 13.1).[9] This temple may have been the forward command post for the NVA division fighting for Hue. After all, the enemy's Tri-Thien-Hue Military Region headquarters was located atop Chi Voi Mountain — about twelve kilometers south-southwest of the Citadel.[10]

Lieutenant Colonel Pham Van Khoa, the South Vietnamese Thua Thien Province chief, remained in hiding until rescued by [the] American Marines . . . [and] overheard a conversation among some enemy officers. According to Khoa, the North Vietnamese mentioned a division taking part in the battle and the division headquarters was "in an unknown location south of the city of Hue inside a pagoda."11

> — *U.S. Marines in Vietnam: The Defining Year, 1968*
> History and Museums Division, HQMC

## The Eastern Corridor

About 4 March, 2/5 made another interesting discovery to the east of the city.

> Leaving the southern sector to the 1st Brigade, 101st Airborne on the 29th, the two Marine battalions [2/5 and 1/5] entered their new area of operations to cut off any NVA forces trying to make their way from Hue to the coast. Although encountering few enemy forces, the two battalions uncovered "fresh trench work along the route of advance, 3000 meters long with 600 fighting holes." Captain Michael Downs, the Company F Commander, remembered a trench complex that "traveled in excess of five miles" with overhead cover every 15 meters. As Downs remarked, "that had to be a way to get significant reinforcements into the city." The search for significant North Vietnamese forces proved fruitless.12

> — *U.S. Marines in Vietnam: The Defining Year, 1968*
> History and Museums Division, HQMC

## Pullout Preparations

The enemy may have begun a piecemeal withdrawal from the Citadel as early as 21 February. Late that night, they hit Allied positions with rockets and mortars — possibly to cover the sound of the first groups to leave. Naturally the Allies responded with plenty of counterfire.

> The NVA opened the February 22 fighting in the Citadel with a mortar barrage at 0330 [3:30 A.M.] that targeted Bravo/1/5 and Charlie/1/5. At least twenty 82[mm] mortar rounds fell on Marine positions. . . . The Marines responded with their mortars . . .

Also before dawn on February 22, the VNMC [Vietnam-ese Marine Corps] was struck with a vicious 122mm rocket barrage.[13]

On the morning of 22 February, the NVA assaulted the South Vietnamese positions to their north. But were they really trying to leave the Citadel the same way they had entered — through the Huu, Chanh Tay, and An Hoa Gates? If Oriental armies use feints during attacks, they may during retreats as well. Once again, according to the official record, the Allies fell for the ruse:

[O]n the night of the 23rd, . . . [there was] another NVA attempt to break through South Vietnamese defenses in the western sector of the Citadel.[14]
— *U.S. Marines in Vietnam: The Defining Year, 1968*
History and Museums Division, HQMC

But the enemy had two other ways of leaving the ancient fortress. Healthy, unencumbered NVA could have exfiltrated to the south, and those carrying wounded could have filed out to the east.

## The Southern Swim

Central to Oriental military philosophy is the ability to disperse when outnumbered. Just because the foe entered the Citadel in battalion strength does not mean that he departed that way. NVA squads were accustomed to operating under decentralized control and, unlike their Western counterparts, could function as semi-independent maneuver elements in enemy-controlled areas.

NVA soldiers could have reached the southern corridor by swimming the Perfume River in the shadow of its bridges and then crawling through the sewer ditches of southern Hue. With one squad departing the Nha Do Gate every 20 minutes, a whole regiment could have departed the fort in three nights.

## The Eastern Trek

Very possibly, the eastern corridor had been used for the evacuation of enemy wounded throughout the 24-day battle. That would

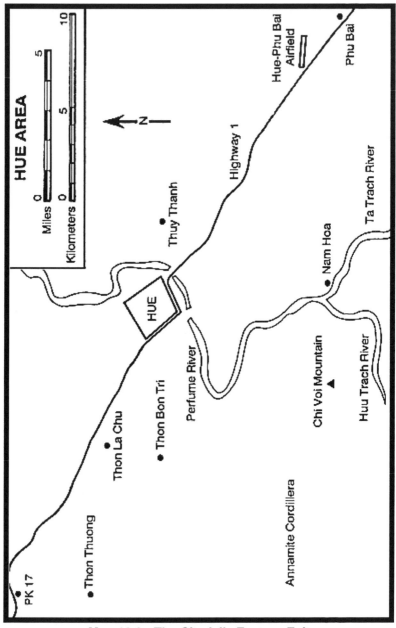

**Map 13.2: The Citadel's Eastern Exit**

(Source: Courtesy of Pacifica Military History, from *Fire in the Streets: The Battle for Hue, Tet 1968*, © 1991 by Eric Hammel)

explain why the enemy had fought so hard for "Phase Line Green." The underground passageway from the Imperial Palace might have run beneath Mai Thuc Loan. Or the street may have had improved, covered sewer ditches. After the Marines crossed over Mai Thuc Loan to attack other streets to the southeast, they did not closely watch it.

The final entry in a captured NVA document — "Twenty-Five Days and Nights of Continuous Fighting and the Wonderful Victory" — may hold other clues to the final pullout to the eastern trench. Could VC sappers in Allied uniforms have helped?

Feb. 23, counterattack and kill 50 enemy, destroy one assault boat, two other boats, kill 23 Americans in Thuong Tu Gate, Dong Ba Gate; capture 25 rifles; during 3 days (21-23) kill 400 Americans at An Hoa (outside northeast corner) [gate on northwestern side]; shoot and burn one helicopter.[15]

It has already been noted that the Citadel walls were laced with Japanese tunnels. Who's to say that a secret passageway did not run the length of the wall between the Nha Do and Thuong Tu Gates, or between the Thuong Tu and Dong Ba Gates? That would explain why the Dong Ba Tower had been so easy for the NVA to counterattack earlier in the battle. On 23 February, 1/5 Marines found heavy tunneling in the southeast wall.

Early in the afternoon, Delta/1/5 was attacking to the southwest, mainly clearing NVA-held bunkers and passages within the southeast wall. At length the company's lead element ran into a large NVA force holed up in a series of interconnecting, mutually supporting bunkers and pillboxes.[16]

Whether a secret tunnel ran the length of the southeast wall doesn't really matter; its base remained in enemy hands throughout the night of 22 February.[17] Its top was not captured until after dawn on the 24th.[18] If the night of the 23rd was dark, a large enemy force could have moved unseen around the eastern corner of the Citadel by simply following the river bank. As has already been shown, the VC still controlled the eastern approaches to the fort.

**193**

Either way, during the period of 21-26 February, the region east of the Perfume River remained free of Allied interference. The two ARVN battalions initially operating in that area had been evicted by a large enemy force during the first few days of February.[19] That enemy force probably had as its mission to build and protect the evacuation route. The two ARVN Ranger battalions assigned to later clearing that zone did not arrive in Hue City until 22 February,[20] and had not captured the built-up area between the northeast wall and the river until 25 February.[21] In other words, any NVA unit that could have reached the road outside the Dong Ba Gate before then was probably home free. The VC infrastructure could have helped that unit to move unseen to the covered trench. Only required would have been porters and guides. The need for boats could have been eliminated by a bamboo bridge just below the surface of the Perfume River.

## Rear-Guard Action

While the Imperial Palace and Huu Gate were captured on the 24th of February, the extreme southern corner of the Citadel did not fall until early on the 25th.[22] The rear guard may have gone east or south, or simply hidden in secret, well-provisioned, underground rooms. After a week or two of laying low, they could have easily blended in with the local populace. In addition to 140,000 inhabitants, Hue had Tet pilgrims.

Throughout the Vietnam War, most of the countryside outside built-up areas belonged to the local VC at night. By the time the NVA rear guard left the Citadel, Hue City's "underground railway" had been well oiled.

## The Defenders Had Achieved their Strategic Goal

An estimated 5000 enemy soldiers had entered Hue City.[23] As the Allied cordon tightened, those south of the Perfume River moved into the Citadel. Before they mysteriously withdrew, they did a lot of damage. By their own estimates, they incapacitated 3630 Allied soldiers, 23 boats, 8 APCs, and 50 aircraft (most parked at Tay Loc Airfield).[24] They also destroyed many Allied tanks.

On the first day of the battle, the [7th Armored Cavalry]
unit commander, Lt.Col. Phan Huu Chi, barely got out of
his own compound gate [about a mile down Route 1 from
the Citadel] — when his column of 26 M41 light tanks and
a dozen M113 APCs were *[sic]* ambushed before they had a
chance to get going. Enemy forces turned back the armored
column with a barrage of B-40 rockets and heavy
machinegun fire. Several of Colonel Chi's tanks were captured by enemy forces, which later used the 26-ton vehicles
against the U.S. Marines. One of the tanks actually made
it across the Perfume River.[25]

Disturbingly, the NVA had claimed less damage to Allied troops
than the Allies would later acknowledge themselves.

[U.S.] Marine units of Task Force X-Ray sustained casualties of 142 dead and close to 1,100 wounded (1/5 AAR Hue
City; 2/5 ComdC, Feb68; 1/1 ComdC, Feb68). U.S. advisors
with the 1st ARVN Division in Hue reported 333 South Vietnamese Army troops killed, 1,773 wounded, and 30 missing in action. According to the U.S. Marine advisors with
the Vietnamese Marine task force in Hue, the Vietnamese
Marines suffered 88 killed, 350 wounded, and 1 missing in
action. The 1st [U.S.] Cavalry Division (Airmobile) listed
casualties of 68 killed and 453 wounded in its battle account. Thus, all told, allied unit casualties totaled more
than 600 dead and nearly 3,800 wounded and missing [4400
people in all]. . . . According to the South Vietnamese, captured Communist documents admitted to 1,042 [enemy]
killed and an undisclosed number of wounded (TF X-Ray
AAR Hue City; 1st InfDiv, Adv Tm 3, CAAR, Hue; Budd,
AAR; 14th MHD, "The Battle of Hue," Mar68; Pham Van
Son, *The Tet Offensive,* p. 271).[26]
— U.S. Marines in Vietnam: The Defining Year, 1968<br>History and Museums Division, HQMC

With less refined evacuation procedures, the NVA often reported
equal numbers of killed and wounded. If 5000 soldiers had been
engaged, 1000 killed, and 1000 wounded, then 3000 had sneaked
away from a besieged fort with many of their wounded. That may
be somewhat of a precedent in the annals of tactical withdrawal.

Perhaps the North Vietnamese never intended to occupy the Citadel indefinitely. The poorly manned 1st ARVN Division headquarters compound in the northern corner was not reinforced until 1 February (a day and a half after it first came under attack).[27] The enemy could have taken it easily but chose not to. Perhaps he only wanted to prove to the American people that the Western style of war was no longer viable — that it had no chance of prevailing in Vietnam. Public resolve cannot compensate for an outmoded style of fighting; it can only extend the period of bloodshed.

This alternative way of war can be practiced without the political subversion and summary executions that overshadowed the tactical accomplishments at Hue. Before journeying again into the sordid world of war, the American public should demand, of its military, the finer points of the alternative method.

### When Too Many Enemy Units Vanish

There is always the outside chance that all 3000 NVA soldiers hid beneath the streets of Hue until the Americans left. A few at a time, they could have easily melted back into the flood of returning refugees.

The underground delay option is not the product of an overactive imagination. In fact, it constitutes an integral part of the North Vietnamese plan to thwart the next Chinese invasion.

A new concept has been devised to meet this contingency — called the Military Fortress. While new and innovative, it does have roots in the "combat village" of the Vietnam War and the still earlier "fortified village" of the Viet Minh War.

The Military Fortress concept presently involves some two dozen [North Vietnamese] districts that abut on China — an inaccessible region of mountain, jungle, and Montagnard — that are to be welded into one contiguous defensive structure. Each village of the district is to become a "combat village," linked in tactical planning terms to neighboring villages; the entire district thus becomes a single strategic entity, and all the districts together become a grand Military Fortress. Villagers are armed, and all have combat duties. . . . Each villager spends part of each day training and working on fortifications, for which he gets extra rations. The work includes digging the usual combat trench foxhole, trench, bunker, underground food and weapons storeroom, and the *ever-present "vanish underground" installation,* the hidden tunnel complex [italics added]. These are within the village. Some distance out, usually two to three kilometers, is what is called the "distant fortification," a second string of interlocked trenches, ambush bunkers, manned by well-equipped paramilitary troops serving full time. Several villages (usually about

(Source: *FM 21-76* (1957), p. 71)

five) are tied together by communication systems and fields of fire into "combat clusters" (about seven per district), and the whole becomes a single strategic entity.[28]

As there is every indication that Eastern forces learn from each other, U.S. forces may very well see again what this part of the book has discussed.

# Part Three

## The Next Disappearing Act

Those who understand big and small units
will be victorious. — Sun Tzu

# 14 How Much Has War Changed?

- How does technology affect small-unit infantry tactics?
- Will modern firepower restrict a "low-tech" Eastern army?

(Source: Courtesy of Cassell PLC, from *World Army Uniforms since 1939*, © 1975, 1980, 1981, 1983 by Blandford Press Ltd.; Corel Gallery, *Clipart* — Plants, 35A022)

## Established Differences

Many Americans have come to believe that the only way to enhance militarily preparedness is through more modern equipment. Since countering the machinegun with the tank in WWI, the West has depended almost entirely on its technological advantage to keep pace with the evolution of war. The only Western nation to counter increased weapon lethality with tactical innovation has been the Germans.[1]

While U.S. military planners generally agree that the machinegun has changed warfare forever, many fail to appreciate why. Since its introduction, interlocking bands of continuous, grazing fire have made moving across contested ground like dodging the blades of an electric meat slicer. Any crawl spaces beneath the fire can be covered by mine or preregistered mortar. It doesn't matter whether the gunners can see their targets. The ground has been walked, sectors of fire established, and guns aimed by search and traverse mechanism. It makes no difference whether opposition supporting arms plaster the area. The gunners have retreated far below ground and then climbed back out to shoot from tiny, concealed bunker apertures. Since the invention of the machinegun, moving through enemy territory has been extremely dangerous. Why Western forces have been slow to acknowledge the danger is anyone's guess. Its terrible toll can be easily balanced against enemy casualty estimates.

> By the end of April [1917] more than a quarter of a million French casualties had been suffered for a gain of some 500 yards of ground [during the Chemin des Dames Offensive].[2]

While U.S. commanders still see the machinegun as a force to be reckoned with, most assume that their parent organization's small-unit infantry tactics have long ago been adjusted. They prefer not "to dwell on past wars," but rather "to get ready for the next one." They assume that the new surveillance and targeting devices will make the lessons of history irrelevant — that (1) the older weapon systems can no longer hurt them, and (2) that any failure to adopt or retain countertactics will no longer make any difference. They expect future opponents to do the following: (1) stay in the open, above ground, and relatively motionless; (2) operate at company size or larger; and (3) closely control their subordinates. In other words, they expect their opponent to fight the way they do and have soldiers of commensurate ability. This chapter will investigate to what extent the new technologies might restrict a low-tech Oriental army.

## America's Current Blueprint for Change

After continuing to operate in much the same way since WWI, the U.S. military has now acknowledged the need for change. It has identified the problem as being with its light infantry forces.

[O]ur light forces lack the staying power that we need. . . .
Light forces must be more lethal, survivable, and more tac-
tically mobile.3
— Chief of Staff, U.S. Army
*Frontline,* 24 October 2000

Unfortunately, the U.S. Army plans to solve this problem with
another piece of equipment — to give its infantry forces the additional firepower and mobility of a Light Armored Vehicle (LAV). It
can now send a "medium" brigade of America's youth into harm's
way within 96 hours.4

## America's "Precision Age" Focus

To minimize friendly fire losses, U.S. commanders have always
kept close track of their subordinate units. The new Global Positioning System (GPS) will greatly facilitate this process. Now any
subordinate-unit leader can determine his exact location through
the resection of satellite signals. In theory, he needs only a laser
rangefinder and electronic compass to determine the whereabouts
of any antagonist. Then, from GPS-equipped artillery batteries,
he can supposedly summon pinpoint fire. Of late, this game of
precise positioning has been extended all the way down to the individual rifleman.

In essence, the U.S. has been trying to digitalize the battlefield
— to merge various pieces of intelligence into a virtual reality and
then to use that frame of reference to orchestrate a war.

The [U.S.] Army learned from its experiments with digita-
lization at the National Training Center in 1997 that a properly internetted maneuver brigade, provided with an immediately available suite of aerial sensors, could expand
its area of control by a factor of four or more. Superior situational awareness allowed units to locate all friendly units
and most of the enemy immediately around them. The ability to see the battlefield with great clarity and immediacy
allowed them to anticipate each enemy movement and avoid
being surprised. Also, units participating in these force-
on-force experiments discovered that the ability to spread
out yet still remain cohesive and able to maneuver freely

allowed them to outflank and surround much larger opponents in open combat.[5]
— Commandant, U.S. Army War College
*Armed Forces Journal International,* December 1999

The Army's reasoning goes something like this — once detected by aerial sensors, an enemy unit will be investigated by scouts, loosely encircled by forces, and then demolished by precision fire.

The scouts' mission would be to define the outline of the enemy formations clearly and then locate and destroy all significant points of resistance without the finding force becoming decisively engaged or suffering casualties.[6]
— Commandant, U.S. Army War College
*Armed Forces Journal International,* December 1999

The Army's plan is not without merit. Oriental armies use a lot of encirclement. Staying beyond decisive-engagement range makes more sense than never backing up. But too little attention may have been paid to the size and tactical ability of the quarry. Encircling an undergunned but otherwise powerful enemy force involves dangerously dividing one's own. Each encircling unit must have the tactical skill to weather a concerted attack. It cannot depend on its firepower alone. Below company size, American units seldom acquire advanced infantry methods.

Although combat experience should indicate otherwise, the rifle squad currently occupies a relatively minor place in Marine Corps tactical thought. Squad level training and doctrine seem to suggest that the squad has little independent tactical value. The squad has been relegated to the role of subunit whose movements are closely controlled by the platoon commander. Considered in terms of maneuver warfare, this attitude is disastrous . . . (maneuver warfare demands that the squad assume a primary tactical role).[7]
— Bill Lind
personal advisor to 29th Marine Commandant

Above company size, American units have also historically exercised more firepower than tactical proficiency.

In my judgment our forces were not as well trained as those of the enemy, especially in the early stages of the fighting. After the buildup of forces, when we went on the offensive, we did not defeat the enemy tactically. We overpowered and overwhelmed our enemies with equipment and fire-power.[8]

— Lt.Gen. Arthur S. Collins, Jr., U.S. Army (Ret.)

This failure to fully practice the art of maneuver has put U.S. forces at a tactical disadvantage in the East. To successfully counterattack the Chinese in Korea, Americans had to ball up and demolish everything in their path.

The advance of the seven American divisions now in the line was the twentieth century successor to the Roman "tortoise": instead of long columns, exposed to surprise attack, Ridgway's units now deployed at every stage for all-around defense in depth, securing themselves against infiltration while they waited for the massed artillery and air strikes to do their work upon the Chinese positions.[9]

The U.S. Army's plan to disperse in the face of modern weaponry is a good one. With highly trained troops, that is the best way to limit casualties. But a commander cannot decrease his unit's structural integrity without increasing its vulnerability to ground attack. If in any way constrained, his separate elements will be easy for the enemy to destroy piecemeal. To survive outside the protective umbrella of the parent unit, his subordinates will need extraordinary skill and to make many of their own decisions. For the plan to succeed, the American Army must do what its Asian counterparts have done — (1) give highest priority to the training of small units and individuals, and (2) decentralize control. Then, though badly outnumbered, each tiny U.S. contingent will have the ability to operate (and survive) semi-independently.

The decentralization of tactical control forced on land forces has been one of the most significant features of modern war. In the confused and often chaotic environment of today, only the smallest groups are likely to keep together, particularly during critical moments.[10]

— John A. English, *On Infantry*

RICE
PADDIES

ELECTRICAL
WIRES TO DEMOLITIONS

(Source: *Counterguerrilla Operations,* FM 90-8 (1986), p. C-22)

Traditionally, the U.S. military has tried to save lives through closer supervision. With the digitalized battlefield, it may be trying to disperse its forces without decentralizing control. During the encirclement of a resourceful enemy, even adequately trained units may have trouble surviving without the leeway to make their own spur-of-the-moment decisions.

With this plan, the Army makes some dangerous assumptions about the enemy: (1) that he has just arrived, (2) that he will allow himself to be reconnoitered, and (3) that he will stay long enough to be shelled. What if he routinely employs feints? Could he not create so many fictitious targets as to overload the American supporting-arms capability? Or what if the enemy is within crawling distance of a bombproof shelter? Could he not monopolize all available U.S. firepower? Too little attention may have been paid to the potential dichotomy between unlimited enemy dispersion and limited American response.

An Eastern foe is not going to passively sit while U.S. forces encircle and systematically destroy him. And where in the U.S. inventory are scouts that can get that close to a Japanese, Chinese, or North Vietnamese formation. Those on foot would have to elude roving sentries who look like bushes, track like Tonto, and smell ivory soap at a quarter of a mile. Motorized scouts would experience a slightly different problem. As their last memory, they would see a lone RPG gunner rising up out of a roadside spider hole. The "Age of Precision" favors the army with the best small-unit and individual skills.

> The Gulf War was the last of the great machine age wars.[11]
> — Commandant, U.S. Army War College
> *Frontline,* 24 October 2000

To the uninitiated, loosely encircling and then crushing an opponent with smart munitions may sound vaguely plausible. But the idea is blatantly attritionist. Wars are won by eliminating targets of strategic import, not killing people. More appropriate might be hundreds of tiny maneuver elements. Eastern squads can acquire enough skill to operate semi-independently. Why couldn't better-trained U.S. squads take turns hiding, bypassing enemy formations, and shelling opposition convoys, ammunition dumps, and fuel farms?

In essence, the new strategy is just a spruced-up version of an old theme — looking more inward than outward to avoid embarrassing mistakes. It places more emphasis on friendly positions than on enemy maneuvers. Against an adversary skilled at deception and tactical withdrawal, it would only work in virtual reality. The encirclement might occur. And its participants might successfully attack. But the enemy would be long gone, just as he has been since the 1st and 3rd U.S. Armies closed their pincers on a mostly empty salient at Houffalize during the Battle of the Bulge.[12] In the old days, before the cordon could be fully formed, the rear guard of a company-sized Eastern unit would keep U.S. observers' heads down while the main body moved through an unoccupied draw. Under aerial sensors, the main body would simply disperse and move through the loosely formed cordon at several different locations. Most of the subordinate elements would not even come under attack by U.S. supporting arms. Their profusion and proximity to U.S. forces would save them. At first, there would be too many targets to shoot at. Then, the targets would be too close to risk a mistake. In the unlikely event that an Eastern unit of any size became fully entrapped, it could simply hug its closest opponent until dark and then exfiltrate the cordon at other locations. Two-man sapper teams can go through Western-style defenses like water through a fish net.[13]

## America's Aerial Sensors

Senior U.S. commanders now have a safe way to quickly assess the power of their opposition. With a specially equipped Remotely Piloted Vehicle (RPV), they can closely inspect unobstructed terrain, day or night. They may soon be able to do the same thing through satellite link-up. With enough of these "all-hours eyes in the sky," they might be tempted to conclude that future wars can be structured.

One wonders to what extent these new surveillance devices would restrict a potential adversary's ability to hide. Certainly, the U.S. does not have the wherewithal to simultaneously monitor each square foot of every battlefield. Furthermore, the most advanced machine will still have limitations. There are many things through which the new surveillance equipment can't see: (1) triple canopy vegetation, (2) soil, and (3) roofing materials, just to name a few. A closely watched foe could still operate: (1) below ground, (2) beneath dense foliage, (3) inside buildings, (4) adjacent to or inside U.S. controlled areas, (5) within a civilian populace, and (6) while too widely dispersed to target.

How much the new equipment has changed combat is evident from "War on Drugs" statistics. Despite a yearly budget of $20 billion and full military cooperation, the U.S. "drug administration" has failed to put a dent in the world's biggest industry.[14] If paramilitary gangs can elude modern surveillance, professional infantry units can too. Better detecting small contingents of enemy does not necessarily mean that they can be more easily eliminated.

## America's Ability to Penetrate Darkness and Obscurants

Advances in U.S. surveillance and targeting equipment have occurred so rapidly as to confuse the best military minds. Fortunately, their underlying concepts are not that difficult to understand. Electromagnetic energy takes many forms. In order of increasing wave length, it covers the following spectrum: gamma, X-ray, ultraviolet, visible, infrared, microwaves, and radio waves.[15] The new surveillance and targeting devices generally fall into the last four categories: (1) those that magnify reflected, visible light; (2) those that sense natural or reflected infrared light; (3) those that detect reflected microwaves; and (4) those that pick up reflected radio waves (conventional radar). All four categories can penetrate the night, but experience varying degrees of difficulty with obstacles, rain, fog, mist, smoke, dust, and haze.[16] To make good sense of the new gadgetry, one must only realize that any heat source emits infrared light.[17]

*Image Intensification Devices (Night Vision Goggles)*

Night Vision Goggles (NVGs) need very little ambient light to operate. The latest versions can see well under cloud cover and no moon.[18] The additional shadowing of a jungle canopy might give them problems. They cannot see through bushes. They cannot see through battlefield obscurants — smoke, dust, haze, or adverse

(Source: *FM 7-8* (1984), p. G-4)

weather — or be used in the daytime.[19] One of the best ways to defeat NVGs during a night attack is to run a continuous illumination mission a mile or so behind the assault force.

Some versions of NVGs cover only one eye, theoretically permitting the other eye to retain its night vision and depth perception.[20] Assumed, of course, is the wearer's ability to concentrate on two things at once.

*Thermal-Imaging (Infrared Sensor) Devices*

Forward-Looking Infrared (FLIR) systems can detect the heat emitted by human beings and combustion engines at considerable distances. Unlike NVGs, they can see through some obscurants and be used in the daytime.[21] But they have trouble seeing through water vapor and precipitation. To them, individual drops or flakes look black. In dense fog, they will not work at all. An "ocean haze" was recently enough to completely obscure a drug boat being chased by a thermal-imaging-equipped U.S. Coast Guard helicopter north of Cuba.[22] A human being can avoid detection under three other circumstances: (1) when the rain sufficiently cools his skin, (2) where ground and body temperatures have equalized at dawn or dusk, and (3) while wearing heat-defecting clothing. Additionally, thermal-imaging devices lack the resolution of NVGs at short distances.[23]

There will soon be an infrared sensor small enough to mount on the U.S. service rifle. Discussed in the next chapter will be its durability to the rigors of prolonged combat.

*Ground Surveillance Radar*

Ground surveillance radar has been around for a long time. On the battlefield, it has often been used to detect motion. The word "radar" comes from the expression "radio detection and ranging." Early models consisted of a radio transmitter and receiver. They transmitted radio signals across a wide area. Signals that "bounced" would provide a target azimuth.

Most detection and ranging devices emit short bursts of electromagnetic energy called pulses. Pulse frequency is set to give each echo time to get back to the receiver before the next pulse is emitted. Pulse width depends on minimum range. Modern radar equivalents use energy of shorter wave length.

Described in the next chapter will be a hand-held, microwave version that can detect a human being right behind a masonry wall.

## America's Ability to Pinpoint an Enemy's Position

U.S. forces now have the technology to precisely locate a clearly visible, hardened target. It should help them to hit such a target.

Unfortunately, the new technology works less well against a partially hidden, soft target — like a widely dispersed infantry formation. In the attempt to pinpoint enemy forces, too little attention may have been paid to whether those forces are on the move, dug in, or feigning a larger presence. If there is a way to confuse the equipment, the enemy will find it. Too many confirmed sightings could be as disconcerting as none at all.

### Laser Rangefinders

A laser is nothing more than an intense beam of light. Its beam is so narrow that it can project a small spot of light onto an enemy tank or building several miles away. This makes lasers useful for determining the range to those kinds of targets.

On the digitalized battlefield, known observer's location, target azimuth, and target distance produce precise target location. In theory, if everyone uses a GPS, electronic compass, and a laser rangefinder, the days of adjusting indirect-fire are over.

When used as a rangefinder, the laser sends out pulses of light and receives back their reflection. By electronically measuring the time needed for each pulse to make the round trip, the distance to the target can be quite accurately calculated.[24] But most laser rangefinders are, in effect, nothing more than infrared radars. They use IR pulse echoes to determine target distance. Ground-hugging, quick moving, and dispersed infantry units lack the substance to reflect a pulse.

Now, the U.S. military wants a tiny laser rangefinder on every rifle. It may be trying to do more than just improve marksmanship. By giving each soldier a rangefinder and means of communication, it may hope to uncover and precisely locate every enemy on the battlefield. Even if that were possible, every enemy sighting will not be deserving of, or vulnerable to, a supporting-arms response.

### Laser Target Designators

The first two letters of the acronym "laser" stand for light amplification. This light can be either visible or invisible. Of course, the invisible part of the spectrum better preserves the element of surprise. The high-energy lasers are not the red or green variety of science fiction, but invisible infrared ones.[25] To surreptitiously "designate" (mark) a target, a soldier has only to point an infrared laser at it. The light reflected from the target can then be "seen" by the infrared sensors in a laser-guided artillery shell, missile, or bomb.

The larger laser designators bathe hardened, point targets in enough infrared light so that "smart" (optically guided) shells or bombs can home in on them. Their continued applicability to modern combat will depend on the extent to which the foe becomes motorized. If the enemy continues to attack by sapper and resupply by porter (as he has in Asia since WWII), the larger laser designators will only have utility against bunkers and buildings.

Now available is a smaller laser-emission device for mounting on any direct-fire weapon. It projects, onto a prospective target, an infrared aiming point for the weapon. In other words, its infrared beam designates the point of impact for the weapon's "dumb" bullets. This infrared dot can be clearly seen through the weapon operator's infrared-sensing goggles. Unfortunately, all laser target designators require a lot of zeroing.[26] Without enough zeroing, they can throw off the gunner's aim. Additionally, every type of laser target designator may see through dust and smoke less well than claimed by its manufacturer.

> Of significant note, obscuration from dust and smoke on the battlefield proved to be far more challenging than anticipated. It nullified IR lasers and made it difficult for support by fire positions to track maneuver elements.[27]
> — Marine Corps Gazette, September 2000

*Identification, Friend or Foe (IFF) Gear*

The ability to distinguish friend from foe has been the long-time dream of professional soldiers. In "virtual reality," that seems possible. Friendly troops can be fitted with "beacons" emitting radiation near the IR blackbody heat range.[28] U.S. Marines wear the "firefly," a nine-volt, battery-run IR beacon. Unfortunately, the firefly is too small for the IR sensors on U.S. aircraft to easily detect.[29]

The field expedient to an IR beacon is an IR "chem-light." Friendly troops with NVGs have no trouble seeing it.[30] Of course, neither does the opposition. The enemy would also have little trouble mimicking any IR tag.

## America's Land Navigation Capability

U.S. troops no longer practice terrain association (using lineal terrain features like streets on a road map) like they did in the late 1980's. For their whereabouts, they depend on fragile GPS devices — equipment that has been proven not to work where tree branches or man-made structures interrupt satellite signals.

With technology that can see ground contours beneath cloud and tree cover (probably satellite radar), the United States is reputedly remapping the world. While the new process may make more terrain detail available, only a small percentage will fit on a 1:50,000 map. The extent to which "map error" will continue to bother land navigators will depend on the angle from which the aerial images are captured and the way in which they are pieced together. If there remains any possibility of image distortion or human error, the new maps, whether paper or computerized, will not precisely replicate what exists on the ground.

## America's First-Round-Hit Capability

U.S. planners have apparently concluded that, where infantry units and artillery batteries universally establish their locations by GPS, every first round will be on target. "Being on target" has a lot to do with what the target is. A first round might slightly bother an above-ground, area target, but not a hardened, point target.

On the latter, it takes bracketing and continuous precision fire to create enough probability of a direct hit. Wind, climatic conditions, and ordnance variations alone will create a 50-meter margin of error. Neutralizing a modern main-battle tank or well-constructed bunker takes putting a round directly through its skin.

What if the hardened target is moving in a zigzag pattern? Can its location when the shell gets there be accurately predicted? The distance to an erratically moving and repeatedly obscured target is difficult to ascertain by laser rangefinder. Even a stationary, point target would have to be on perfectly level and unobstructed ground, the elevation of which was precisely known. Big trees or structures, small hills or depressions, and slight differences in elevation will significantly alter the point of impact for a low-trajectory round. Because Asians like to dig in, they can only be hurt by direct hits from artillery or mortar fire.

Historically, the problem has been getting fire support in a timely manner. While the digitalized battlefield may more quickly establish whether there are friendly forces near the target, it cannot give every mission the same priority. If every "cleared" enemy sighting deserved an artillery barrage or airstrike, no amount of ordnance would be enough. On an active battlefield, there may be thousands of separate engagements going on at the same time. To whatever extent the U.S. can speed up the bombardment process through electronics, the enemy can slow it down through dispersing his troops and creating fictitious targets.

Eastern forces do not normally announce their presence until their adversaries are close-by. It will be many years before casualty-conscious U.S. commanders will feel comfortable enough with the "Age of Precision" to grant so much as a mortar clearance within 100 meters of friendly troops. To fire safely in the heat of battle, American supporting-arms observers should continue to call their first rounds beyond danger-close range (600 meters for artillery). Otherwise, they may get to relearn the lesson of the Vietnam-era short round.

## Opposition Weaponry

The small arms of the future will be laser weapons.[31] While conventional bullets create a tremendous logistical burden, laser beams have no weight. Future battlefields will be swept by closely spaced pulses (or steady streams) of lethal electromagnetic energy. To move through an area covered by such a weapon, one must do what he would have done with a machinegun — operate in small groups and use microterrain for cover.

To counter the weapons of the future, the U.S. has only to adopt post-machinegun, small-unit tactics. It has only to field squad-sized or smaller units with enough skill to operate alone — as phantoms in every circumstance.

### Still Required Will Be Outsmarting One's Opponent

The problem in Vietnam was not one of too little resolve, leeway, or wherewithal; it was one of insufficient appreciation for how the enemy fought. Until that ongoing oversight is corrected, no amount of technology will be enough. Eastern tactical technique must be studied until each ordinary-forces' feint can be linked to an extraordinary-forces' maneuver. The digitalized battlefield will only draw the U.S. commander's attention away from sapper teams and stormtrooper squads. It will make him easier to mislead.

Moreover, the digitalized battlefield may further erode the small-unit leader's capacity for tactical decision making. With access to the precise locations of every friendly unit and enemy sighting for miles, he will be less likely to suspect hidden danger nearby. Encouraged to focus on the "big picture," he will see little reason to evaluate or exploit local circumstances. To look good on his parent unit's monitor, he will concentrate more on his place in formation than enemy intentions. Tomorrow's small-unit leader may exhibit even less initiative than his father in Vietnam.

### The Oriental Defense Will Still Work

If Marines equipped with all of the latest gadgetry were to try to refight the battle of Iwo Jima, they would have as much trouble as their grandfathers did. Most of the real estate was defended from elsewhere. The Japanese stayed below ground, took turns shooting from tiny apertures, and then withdrew. There is no existing surveillance equipment that can identify as such, a dummy position or below-ground fortification. The preparatory fires would still land in the wrong place. Contemporary defenders could mask their building of a gun emplacement by tunneling or simply blanketing the island with hundreds of fake working parties. During the invasion and ensuing battle, they could still keep their intentions secret by staying below ground. Today's satellites would see a battlefield just as empty as the airplanes saw in 1945.

More recently in Kosovo, the Serbs had little trouble eluding sophisticated surveillance and targeting efforts.[32] The U.S. government recently disclosed that its initial estimates of damage done to targets of strategic importance had been excessively optimistic.

> According to a suppressed Air Force report obtained by Newsweek, the number of targets verifiably destroyed was a tiny fraction of those claimed: 14 tanks, not 120; 18 armored personnel carriers, not 220; 20 artillery pieces, not 450. Out of 744 "confirmed" strikes by NATO pilots during the [78-day] war, the Air Force investigators, who spent weeks combing Kosovo by helicopter and by foot, found evidence of just 58.[33]
>
> — *Newsweek,* 15 May 2000

Where the Serbians learned to weather the onslaught should come as no surprise. Eastern Europeans appreciate the extent to which Asian thought has influenced war.

> "The Art of War" has had a profound influence throughout Chinese history and on Japanese thought; it is the source of Mao Tse-tung's strategic theories and of the tactical doctrine of the Chinese armies. Through the Mongol-Tartars, Sun Tzu's ideas were transmitted to Russia and became a substantial part of her Oriental heritage. "The Art of War" is thus required reading for those who hope to gain a further understanding of the grand strategy of these two countries today.[34]
>
> — B.Gen. Samuel B. Griffith USMC (Ret.)

The Serb reaction to firepower-dominant assault was remarkably similar to that of the North Vietnamese over a quarter century before. The Vietnamese [had] realized that overwhelming firepower alone could never compensate for the presence of an aggressive force on the ground to find, fix, and fight them in close combat. Without a ground threat, they merely had to array their forces in order to endure punishment by fire alone.

Serb tactics followed the North Vietnamese example with remarkable fidelity. Units went to ground and dispersed over wide areas. Soldiers hid their equipment with great skill and constructed dummies that proved effective at spoofing aerial observers and image interpreters.[35]

— Commandant, U.S. Army War College
*Armed Forces Journal International*

## "Ordinary" Oriental Forces Could Still Attack

In some ways, the new surveillance devices facilitate the ordinary forces' feint. As at Khe Sanh, those forces could appear to be converging for an attack while staying well enough dispersed to weather bombardment.

If their mission were to close with U.S. lines, they could create so many small, quickly moving targets as to make no particular one worthy of indirect fire or susceptible to a first-round hit. They would only have to send a score of lone squads in from different directions. If all came in at once without stopping, most would make it. American observers would be unable to change their target locations quickly enough or engage every target. If those squads took turns moving and hiding under heavy foliage or rooftops, they could cover the electronic battlefield display with enough separate readings to adequately confuse any interpreter. Once they were within 500 yards of their objective, safety-conscious U.S. commanders would be hesitant to target them with supporting arms. Just by staying in the draws and ditches, the enemy squads could withstand most forms of direct fire. Only automatic grenade launchers and mortars would give them much grief. Most U.S. positions would not have the assets to handle multiple threats from numerous directions. The enemy squads that came under fire could simply stop and take cover while the others continued forward.

## "Extraordinary" Oriental Forces Could Still Operate

With the defenders' attention fully monopolized by a more visible feint, the extraordinary forces could operate pretty much as they always have. They could do any one of the following: (1) hide below ground during U.S. base construction; (2) tunnel; (3) infiltrate; or (4) simulate/follow a returning U.S. patrol or listening post. Many of the scores of Japanese snipers inside the Henderson Field perimeter on Guadalcanal may have hidden before the landing.[36]

If the battle for Khe Sanh had continued, U.S. forces might have gotten the opportunity to experience the tunneling option. Corporal Bob OBday, a frontline machinegunner, clearly heard digging under his position on more than one occasion. Later, while sweeping the area 200 yards beyond the protective wire, his unit found a 800-square-foot layer of fresh dirt.[37]

Stormtroopers would still have a viable attack sequence: (1) crawl one or two at a time during the day to a location just outside U.S. lines; (2) hide below ground, foliage, rooftops, or heat-reflecting blankets until dark; and then (3) attack. Upon rendezvousing, they would be too close to U.S. lines to hit with indirect fire. Their

(Source: Courtesy of Cassell PLC, from *World Army Uniforms since 1939*, © 1975, 1980, 1981, 1983 by Blandford Press Ltd.)

assault technique would require less than 30 seconds to execute. When ready to attack, they could defeat NVGs with preplanned, well-placed illumination. To ground-mounted thermal imaging, their single column would look like a foraging animal or lone scout, unworthy of final protective fire. In torrential rain or heavy fog, that column might make no image at all.

Thermal imaging would also have trouble identifying a partially submerged sapper. He could still enter U.S. lines through any swampy sector. Where the wet area only bordered those lines, he would remain invisible to thermal imaging for a few seconds after emerging from the water[38] — just long enough to make it to cover. Only if that cover consisted of a shallow ditch, narrow tree, or rock, would he even create an aura.

To aerial imagery, a two-man sapper team would look like a returning listening post, and an enemy squad like an inbound U.S. patrol. A single sapper could still enter a closely monitored sector of U.S. lines by doing the following: (1) during the day crawl to a gate in the barbed wire, (2) cover himself with a heat-reflecting blanket, and (3) after dark fall in behind a returning U.S. patrol or sentry post. Something similar was witnessed by a U.S. Army unit at the Chosin Reservoir.[39] Once inside the position, the sapper could put on a U.S. uniform. The Chinese also used this ruse in Korea.[40] To disappear from view in the 21st century, any enemy soldier would only need a stolen or counterfeit IR tag.

(Source: Gallery Graphics, Mac/EPS, *Flowers, Trees, and Plants 2,* "Shrubs")

## The Foe Needs Only Dispersion and Initiative to Persevere

It has been assumed by many that recent advances in the technologies of night surveillance and precision targeting have finally removed the enemy's favorite operating environment, and with it his combat survivability. To some, this means an increase in U.S. capabilities. They may have forgotten that virtual reality will provide the foe with additional opportunities for deception. They may have forgotten that he can operate almost as well in the daytime. Hopefully, they have not forgotten that killing any infrared image not wearing a friendly IR beacon constitutes a breach of moral ethics.

War will continue to be an art for those who stand to gain nothing from calling it a science. The epitome of that art is to win without killing.

# 15 The One-on-One _____ Encounter

- *What equipment will the future U.S. infantryman carry?*
- *How will it help him to defeat his Eastern counterpart?*

(Source: Courtesy of Cassell PLC, from *World Army Uniforms since 1939*, © 1975, 1980, 1981, 1983 by Blandford Press Ltd.; Corel Gallery, *Clipart — Plants*, 35C087)

## The Anatomy of a U.S. Casualty

Americans have been told that two variables determine an infantry unit's casualty rate: (1) the tactical-decision-making ability of its commanding officer, and (2) the quantity/quality of its equipment. They assume that, with sufficient direction and wherewithal, their sons and daughters can win in war without getting hurt. But infantry combat is not like that, it is highly fragmented and chaotic. Every battle consists of hundreds of separate engagements. How few U.S. soldiers get injured in each engagement depends on how quickly they can master unique circumstances. What the commanding officer decides from another location is far less important than what his soldiers and small units have practiced beforehand. And their equipment's potential is far less important than its functionality. Yet, to help the contemporary U.S. soldier to survive his trial by fire, he has been given quicker thinking lieutenants and more complicated equipment.

Saving one's life in combat takes more than just good leadership, it takes knowing how and when to dodge a bullet. The "how" is called individual skill. "When" is dictated by survival instinct. For either ability to do the U.S. soldier any good, he must have the freedom to exercise it. Engagements occur between small groups or individuals. Who wins depends on the relative skill and initiative of the participants. Even with a direct radio link to the subordinate involved, the platoon or company commander can only offer belated advice on incomplete information. For the young rifleman struggling to survive, the leeway to react to unforeseen circumstances can be far more helpful than following every instruction.

The latest equipment best helps he who decides when to use it. How much it might help the U.S. infantry private in a close encounter with his Oriental counterpart deserves more discussion.

## The Latest Attempt to Eliminate Human Error

Well accustomed to Americans' distaste for battlefield casualties, the U.S. Army has come up with a way to reduce its friendly fire losses. In September of 2000, it ran a 6000-man experiment at Fort Polk called "Land Warrior." The future U.S. infantryman will know where he and his buddies are at all times. His headquarters will not only know where each soldier is, but also what he sees.

> The . . . soldiers are equipped with a hand-built digital system that includes everything a soldier wears and carries in the field . . . Each soldier has a tiny personal computer packed in silicon gel strapped to his back. The computer is connected to a thermal sight mounted on the soldier's weapon as well as to a laser range finder, digital compass and video sight.
>
> All the information the equipment collects — grid coordinates, terrain maps, video images and e-mail messages and operations orders — is displayed on a small screen mounted over the soldier's right eye.

The information can be sent to other soldiers or to higher echelons. The display even allows a soldier to shoot his weapon around corners without moving from his covered position.

All of this is meant to allow combat soldiers to communicate with each other, to know their positions always and to precisely pinpoint targets.[1]

## U.S. Soldiers as Information Gatherers and Gun Platforms?

Enthralled with the new equipment's management potential, U.S. Army planners have apparently decided to remove from the equation much of the individual soldier's initiative and God-given senses. With enough tactical skill, that soldier might have been able to compensate for the additional guidance.

**Figure 15.1: Land Warrior**
(Source: Courtesy of Raytheon, © 1999)

Today's U.S. Army infantryman, despite advances in weapon, night-vision, and other technologies, fights much the same way as he did during World War II. The Army's dismounted soldiers still use paper maps, communicate with each other by voice or hand signals, and often lose track of each other as they maneuver on the ground. . . .

The Land Warrior program . . . aims to change all that dramatically. For the first time, the soldier's equipment is being designed as if he is a system — a complete weapons platform . . .

. . . [T]he [L]and Warrior system will give the infantryman significant new capabilities — improving his lethality, mobility, survivability, and awareness of the tactical situation . . .

The heart of the integrated Land Warrior system is a small, wearable, computer-radio subsystem, mounted on the soldier's lower back. The computer is connected to the Thermal Weapons Sight (a Raytheon infrared sensor. . . ) atop his rifle . . . The computer is also linked to a combined laser range finder/digital compass and a video TV camera sight, both mounted on the weapon. The computer displays imagery that the soldier views through a helmet-mounted, monocular viewfinder covering one eye. The Land Warrior soldier thus sees a miniature computer screen — a "head-up" display — that shows digital maps, graphics, and text in a Microsoft Windows, pull-down-menu format, as well as Thermal Weapons Sight or video camera imagery from the direction he points his weapon. . . .

A built-in Global Positioning System (GPS) satellite receiver provides the soldier's location to the computer, which also receives location reports from other soldiers in the squad. . . . The Land Warrior can use the laser rangefinder to pinpoint a new enemy position, which then appears as an icon on all of the squad's map displays.

The "mouse" control for the computer's menu-driven displays is a small button on the side of the weapon that the soldier manipulates using the fingers on his trigger hand. Each soldier, using a helmet-mounted microphone that sits in front of his mouth, can talk with others in his squad via secure voice radio.[2]

—*Armed Forces Journal International,* October 1999

While this gadgetry may appear to make the U.S. rifleman more lethal, he won't decide when to use it. He will be operating as an extension of his headquarters. Through him, his commander will more easily detect and shoot at fully exposed, distant opponents. The individual rifleman will be less aware of what transpires in his immediately vicinity than was his grandfather. With one eye on a computer screen, lips talking through a microphone, ears listening to a speaker, and trigger hand playing with a mouse, he may have a hard time sensing nearby danger. He may not detect the black apparition rising up out of the ground behind him, the slightly wilted foliage on a mound to his direct front, or the faint sound of an AK-47 coming off safe. Then with one eye covered and 90 pounds of extra gear,[3] he may not be able to outmaneuver the person trying to kill him. Red-dot targeting, silenced rifle, and expert marksmanship may do little good where that person refuses to show himself. Only to Americans do weapon performance and marksmanship skill equate to nighttime assault expertise.

> The primary assault weapon carried by most Rangers is the M4A1, compact version [that can be silenced] of the M16. Their marksmanship program is designed to train them to hit what they can see, day or night. In fact, the Army's Rangers are likely the most effective night-assault force in the world. They are equipped with the Army's latest night vision goggles and use a variety of thermal night sights and laser aiming devices on their M4 carbines.[4]
> —*Armed Forces Journal International,* October 1999

The individual weapon of the immediate future will be the Objective Individual Combat Weapon (OICW). It is a combination weapon capable of firing both 20mm high-explosive rounds and conventional 5.56mm kinetic-energy rounds.

> The integrated, electro-optical, full-solution fire control system (FCS) is a first for infantry rifles. The FCS includes a laser rangefinder, optical sight, electronic compass, and air temperature and pressure sensors. . . . The optical sight has a zoom capability. . . . It also has a TV camera with a video processing capability that can be used for individual frame grabbing and recording or for automatic target tracking. The simple red dot day/night sighting [device] uses an

uncooled infrared sensor technology. . . . Images from the infrared (IR) sensor and TV camera can be sent to a helmet-mounted display as well as to other squad members, other units, or higher level commanders via wireless communications links.

Being a full-solution FCS means that when the gunner acquires the target, the laser measures the distance to [the] target. The drag of the projectile is determined using the measured air temperature and pressure and weapon angles. These parameters are used to obtain the ballistics solution that gives a compensated aim point. Before the 20mm round exits the barrel, the FCS "tells" the fuze on the fragmenting antipersonnel round when to fire. . . .

. . . [T]he 20mm ammunition will be effective at ranges up to 1,000 meters . . . While the weapon does not exactly shoot around corners, the shooter can choose a point in space where the bullet will explode — for example, 1 meter after it passes the corner or after it has dropped below a hill or embankment.[5]

— *Marine Corps Gazette,* January 1999

Soon available to the U.S. small-unit leader will be a radar flashlight. It can help him to determine which areas are occupied.

. . . The RADAR Flashlight uses a microwave radar combined with a very fast signal processor to detect minute movement of the human thorax due to the person's heartbeat or breathing. . . .

The RADAR Flashlight can detect respiration of a human behind a wall, door, or in an enclosed space with nonconductive walls at distances of up to about 10 to 16 feet, depending upon the material. It can also detect a person behind foliage, such as bushes, amongst a tree line, or behind a chain link fence. It can detect an individual behind a solid wooden door or standing 4 feet behind an 8-inch block wall.[6]

— *Marine Corps Gazette,* January 1999

Future opponents will not fight in open, unobstructed terrain. They will fight in the mountains, jungles, or cities. That is where their small-unit infantry skill will help them the most.

The Army increasingly sees urban terrain as the battle-field of the future, sought out by enemies who hope to neutralize America's might by getting its forces into a close fight.[7]

— *Frontline,* "The Future of War"

## NCOs Relegated to Equipment Maintenance and Training?

Unlike German, Russian, Japanese, Chinese, North Korean, and North Vietnamese commanders, American officers do not normally solicit the tactical advice of their infantry NCOs. That those NCOs will still offer occasional input has generated more than one attempt to keep them busy doing something else. At last, this leadership problem has a solution. The personal equipment of the individual soldier has become so complicated that it cannot be operated without the full attention of his NCO.

It is the job of the NCO to ensure the junior Marines know how to use the equipment, clean the equipment, and inspect the equipment to see that it is fully operational. Once the NCOs set the standard and show that they do not tolerate Marines going to the field without batteries for their NVGs or without knowing how to set mechanical zero on their PAQ-4 [rifle night-targeting device], then a unit can function as it is designed to do. . . .

The keys to all our attacks in the IR spectrum lay with the NCO's ability to understand, disseminate, and then rehearse the plan prior to execution.[8]

— *Marine Corps Gazette,* September 2000

## The New Gear Creates Opportunities for the Commander

Sharing the new technology with the lower echelons of the organization is certainly a step in the right direction. Over the years, enlisted infantrymen have seldom had the luxury of using high-priced equipment. Unfortunately, the Land Warrior's gear will benefit the chain of command most. His commander will learn from the GPS where he is, and from the camera/rangefinder, what he sees. To make wiser "top-down" decisions, Western-style commanders will always want to know more about what their skirmishers do and see.

It would be nice if the new gear satisfied more of the individual soldier's needs. He would want it to provide more survival options. He realizes that better seeing an opponent in no way guarantees that he has not been seen himself. He understands that knowing an adversary is there in no way guarantees that he can kill that adversary. It can take hundreds of dumb rounds to hit a running human being. There are not enough smart rounds in the world to hurt one who is well enough dug in. At the Somme, a German machinegunner had only to descend the fifty feet into his "dugout" to survive a direct hit from a massive shell. The Land Warrior will only be lethal against a fully exposed enemy soldier who doesn't know the Land Warrior is coming.

With the new equipment, the average rifleman could use more utilization leeway and surveillance options. In a firefight, he will instinctively shed whatever hampers his movement. If an infrared camera can detect sick vegetation from space,[9] couldn't it detect the wilted foliage around an enemy bunker? Highly reflective of near-infrared energy, healthy plants look bright in infrared photographs. Diseased plants look dull and dark.[10]

## The Gear Poses New Problems for the Individual Rifleman

U.S. planners may have mistakenly equated a more lethal rifleman with one who is harder to kill. When matched, one-on-one, against his highly skilled but underequipped Oriental counterpart, he will have no more chance than Goliath against David. The oversight may be one of scale. When the Oriental soldier suddenly appears out of the ground, the U.S. rifleman could care less about what is happening 200 yards away. His world has suddenly shrunk into a 50-meter square. He already knows where his buddies are and cannot employ supporting arms without careful adjustment.

What helps a commander to more easily control his unit may not help his subordinates to more easily survive. In a close-in, life-or-death struggle, the Land Warrior's capabilities will only be enhanced in two ways: (1) shooting around corners at fully exposed targets, and (2) detecting at night fully exposed targets.

Asians will seldom allow themselves to become fully exposed. In the old days, the U.S. infantryman relied on "pieing off" and night vision techniques to accomplish the same things. While the old skills may have been slightly less effective than the new equipment, they didn't increase the soldier's load by 90 pounds.

It is manufacturers who try to ignore product limitations. The individual soldier does not have that luxury.

### Personal Infrared Sensors

The incessant dampness of combat has always created problems for sophisticated equipment. Now the precipitation itself will come into play. Water vapor absorbs infrared rays.[11] In heavy rain or fog, the Land Warrior's infrared sight will not work at all. In good weather, it will not see small objects as sharply as would a telescopic sight. Infrared pictures look "fuzzy."[12] This may affect the Land Warrior's ability to discriminate between targets.

The new infrared rifle sight has other disadvantages. Just to use it effectively, the individual soldier must learn how to change its batteries and mechanically zero it after dark. Doing so rapidly might take some semblance of night vision. It is also uncooled, and as such, has less capability than its larger cousins.

Uncooled FLIRs can't see as far at night but are considerably less expensive and lighter in weight than standard FLIRs, whose detector must be cryogenically cooled to achieve the sensitivity required to distinguish temperature differences.[13]
— Armed Forces Journal International, December 2000

### Night Vision Goggles

The problem with wearing any night vision device has always been one's loss of peripheral and depth perception.

As a Marine moves downrange with night vision goggles (NVGs) strapped to his head, his peripheral vision and depth perception are inhibited. Tasks usually deemed easy by day, such as magazine changes, immediate action, ammo redistribution, and fire and movement, become insurmountable at night without basic, repetitive, muscle memory training.[14]

— *Marine Corps Gazette,* September 2000

## It Will Be Harder for the Land Warrior to Hide

The Land Warrior will be able to detect exposed enemy soldiers at greater distances, not become invisible himself. His new gear will help him to see through unobstructed darkness, not reduce his own nighttime signature. It will in no way diminish his silhouette, noise, or odor. In fact, it will make him even easier to detect in the old-fashioned ways. After 15 years in units that regu-

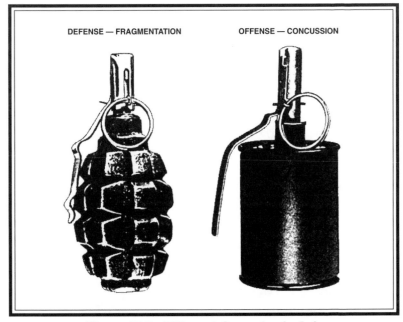

DEFENSE — FRAGMENTATION          OFFENSE — CONCUSSION

**Figure 15.2: Different Grenades in Asia**

(Source: *Handbook on the Chinese Communist Army*, DA Pamphlet 30-51 (1960), U.S. Army, figure 72, p. 74)

larly go to the field, Marine Staff Sergeant Buddy L. Locklear should know. With the naked eye, he has less trouble spotting an aggressor wearing NVGs. The aggressor's need to continually reposition the uncomfortable headgear is what gives him away.[15]

Against the perfectly camouflaged, below-ground defender, such "dead giveaways" will continue to constitute a handicap. Asian patrols are so adept at "reading sign" and tracking that just going the same way twice invites ambush. The new night vision devices may help city-raised Americans to attack each other in training, but they have yet to be proven against woods-wise opponents in combat.

Modern infantry tactics involve little or no firing of small arms. To the soldier under attack, grenades and bazooka rounds sound like long-range mortar fire, but bullets can only mean one thing — a nearby human being. That's why the Chinese use concussion grenades on offense.[16] Initially, the whole idea is to keep defenders from realizing that they are under ground attack. Then, by being the only ones shooting, the defenders give away their positions. Unfortunately, with the Land Warrior initiative, the emphasis has not been shifted from firepower to surprise. The enemy has only to send up an old-fashioned flare to see the solid line of Darth Vaders advancing toward him. The Land Warrior will not be able to watch where he steps or assess his noise. He will have made the ultimate mistake in combat — overestimating his own potential. He will see no reason to stalk from tree to tree and may be far from cover when he needs it.

(Source: Gallery Graphics, Mac/EPS, Flowers, Trees, and Plants 2, "Shrubs")

**231**

## It Will Be Harder for the Land Warrior to Dodge Bullets

It is doubtful that the Land Warrior will be allowed to get down during a nighttime assault. He will be forced to walk upright through interlocking bands of grazing, automatic-weapon fire just as his predecessors were. Even if he were allowed to get down, the new gear would restrict his ability to crawl forward.

Modern infantry assaults involve various combinations of crawling and running (depending on the circumstances). To sustain minimal casualties, assault troops must sneak up on and then quickly close with their opponents.

## The Unpleasant Implication

Throughout history, Western commanders have tried to help their soldiers to survive by more closely controlling them. The Land Warrior initiative appears to be the final step in this well-meaning process. While the new technologies may enhance certain of the individual soldier's sensory abilities, they will reduce others. No longer will a soft click, breaking twig, garlic smell, or sixth sense alert him to impending danger. In peacetime, he will become even more convinced of his immortality. In war, he will be more able to kill a distant, uncovered opponent, but less able to survive one that is nearby and below ground.

The Land Warrior's equipment with blunt his God-given instincts and restrict his freedom. Some commanders may be able to handle the additional power, and some may not. To protect Americans against external evil, the new strategy has made the thinking, caring, and hoping sons and daughters of America into an intermediate stop on the road to robotics. What has been described in the Eastern world as the art of war will be approached in the West as if it were a science — the science of killing. No longer will there be the options of mutual respect or occasional mercy between antagonists. The individual soldier's predisposition toward error will have been effectively removed, and with it his initiative. In the midst of man's most obscene pastime, the conscience of the closest observer will have been effectively overridden.

# America's Only _____ Option

- *What might a future war with an Eastern nation look like?*
- *Has America done all it can to prepare?*

## The U.S. Must Prepare for the Most Probable Threat

As the United States moves into the 21st century, it will continue to have difficulty avoiding armed conflict. Most of its wars will be limited in scope, largely negating its technological advantage.

> [To properly prepare,] it [the U.S. military] is going to have to say [to itself], in order for me to be relevant, I need to specialize in the messy conflicts of the Post-Cold War era.[1]
> — John Hillen
> Commission on National Security

It's doubtful that America will go to war with Britain, France, or one of its allies. Its future adversaries will be from Eastern Europe or Asia. If they fight a different way, then that's the way for which the United States must prepare. If the Eastern style of war has been specifically designed to overcome Western technology, then the new targeting and surveillance equipment will make little difference.

Instead of continuing to discount and ignore the Eastern approach to war, the West must minimally come to understand it. To counter any technological breakthrough, the Easterner needs only smaller contingents and more decoys on both offense and defense. Any increase to the visibility of his ordinary forces will simply draw more attention away from his extraordinary forces. Until the U.S. military develops the capability to see and shoot below ground, Eastern armies will continue to operate with impunity.

The Western focus on technology will seldom be as effective at minimizing casualties and winning wars as the Eastern emphasis on the individual soldier. That soldier's survivability will depend more on him than his commander. His skills will be more durable than his equipment. And his conscience and initiative will be superior to those of a machine.

## Equipment Breaks

The signs are unmistakable. The U.S. Armed Forces fully intend to continue on as before — to try to keep pace with the evolution of war through technological advancement.

> As future uses of image-intensification devices are contemplated, improved night-vision technologies will continue to be the mainstay of the Army's current effort to improve its light infantry forces.[2]
> — *Armed Forces Journal International,* August 1999

When did bullets, shrapnel, concussion, and moisture stop being components of war? Hand-held electronic equipment runs on batteries. It also gets shot, wet, and broken. On Tarawa, a 40-year-old piece of technology failed miserably. The Marines had no radio communications for the first three days.[3] The radios on the ships succumbed to the concussion of the deck guns, and the ones on the beach to the moisture of the amphibious landing. FLIR devices are also susceptible to shock damage. Whether or not a machinegun- or rifle-mounted version could withstand the continual pounding of infantry combat is still very much in question.

> The Army will also retest the medium and heavy TWS [thermal weapon sight] variants at that time [March 2001], Spadafore [the U.S. Army's Night Vision/Reconnaissance, Surveillance, and Target Acquisition Product Director at Fort Belvoir, VA] said, since they suffered reliability problems in operational testing early this year relating to their susceptibility to shock when used with the M4 [carbine] and M240 [machinegun] weapons.[4]
>
> —*Armed Forces Journal International,* December 2000

## Individual Skills Are More Durable

While people make mistakes, practiced skills remain relatively constant. That's why the Eastern commander places such a high priority on keeping his individual soldiers highly trained and totally informed. He realizes that human beings can generally outsmart machines.

While less value may have been attached to the Eastern soldier's life, more value has been attached to his individual contribution. The end product has been fewer Eastern soldiers lost during one-on-one encounters.

> We can still lose this war. . . . The Germans are colder and hungrier than we are, but they fight better.[5]
>
> — General Patton (during the Battle of the Bulge)

## The Hidden Costs of the Western Approach

Historically, U.S. forces have been the best trained in the world — at operating their equipment. Unfortunately, it takes more than knowing how to use one's rifle, radio, and night vision sight to survive in combat.

Of late, American equipment has been upgraded so often that U.S. soldiers have had little time to concentrate on anything else.

Night surveillance and targeting devices have offered an alluring alternative to what few night-fighting skills Americans had. On two recent exercises, U.S. maneuver elements got lost moving toward clearly silhouetted objectives 200 yards away.[6] In effect, new gear has appeared on the scene so often that U.S. troops spend most of their time learning how to operate it.

> Most of [the] Corps' Operating Forces realize that their units lack technical knowledge with regard to night enhancement devices. Entry-level schools do not dedicate the necessary time to teach proper operation. . . . Additionally, technologies have been advancing so rapidly that it is difficult to keep pace in the schoolhouse. Consequently, a dearth of night fighting knowledge exists in many small units within the infantry today. . . .
> . . . The training curve has rapidly fallen behind the equipment development curve, and parity between the two must be reacquired.[7]
> — *Marine Corps Gazette,* September 2000

### The Future of War

The weapons of the future will not be bigger and more destructive bombs. They will be smaller and "smarter" missiles that can recognize their targets. Like the spotting rifle on the old 106mm recoilless gun, an IR beam will designate the point of impact for direct-fire weaponry. Only one shot will be required to kill whatever gets painted with the IR dot.

Tomorrow's military will use laser weapons more powerful and accurate than anything presently in existence. The danger posed by these lasers will change the way wars are fought. Large numbers of soldiers will never be employed together. Wars will be fought by small groups that stay out of sight most of the time.[8] If a target can be seen, it will be hit.

To counterbalance the increase in weapon lethality, military forces must disperse. Both offensive and defensive combat will be conducted by tiny, semi-independent teams of highly skilled soldiers. From now on, being spotted above ground will equate to being unable to move or worse.

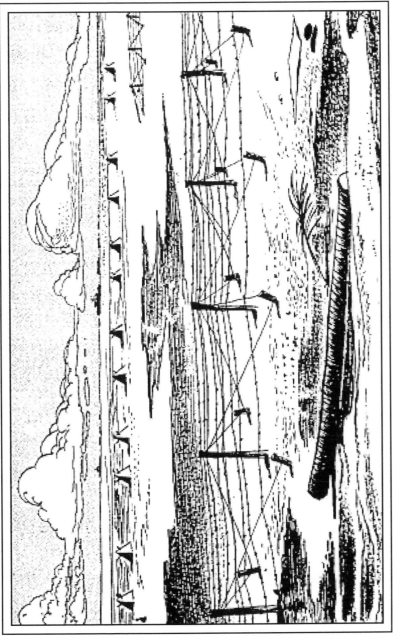

(Source: *Handbook on Japanese Military Forces,* TM-E 30-480 (1944), U.S. War Dept., p. 152)

## Tactical Technique Must Take Precedence over Technology

To assume that technology drives tactical technique is to throw away everything that has been learned about night fighting. In difficult terrain, victory depends more on surprise than on firepower. To win at minimal cost, one must do more than just better see one's opponent — one must keep that opponent from realizing that he is under attack.

U.S. night-fighting techniques have never been as good as those of the Eastern world, and the new equipment has only compounded the problem. Moving upright into an area completely covered by machineguns, mines, and artillery would not work for a fully armored robot.

> Theoretically, aggressive training and tactical innovation should drive technological development.[9]
> — *Marine Corps Gazette,* September 2000

## How Much U.S. Tactics Have Changed

Throughout the 20th century, America's adversaries have initiated most of their offensive actions during the hours of darkness. They know what it takes to stop such an attack. Throughout the same period, U.S. forces have only rarely assaulted at night. With the new night vision devices, they will attempt what they would normally do in the daytime: (1) establish a base of fire; (2) walk on line to their objective; and then (3) assault, killing everyone in their path. Even against an inept enemy, their plan may cost too much. Contemporary Americans will experience additional problems: (1) improper weapon zeroing, (2) low batteries, (3) defective equipment, and (4) lack of operator confidence.

> It takes a company of Marines 4-6 hours to properly zero their PAQ-4Cs [infrared laser targeting devices].[10]
> — *Marine Corps Gazette,* September 2000

Because the Americans' objective will not have been closely reconnoitered ahead of time, this type of maneuver will still constitute a hasty attack against a prepared enemy position. The consequences will be as dire as they have been in the past. A smart enemy does not need to physically occupy ground to defend it. Imagine U.S. Land Warriors moving north on Iwo Jima. They might get away with their stand-up assault one time. Then, the carnage would be worse than it was in 1945.

## Misplaced Trust?

When young Americans join the U.S. Armed Forces, they expect to get more than just the latest equipment. They expect to be told about the small-unit tactics with which every opponent since 1917 has thwarted U.S. firepower. Until that expectation is met, the U.S. military should not talk too loudly about military preparedness. Those trusting, hopeful, mortal, wonderful young men and women have promised to make the ultimate sacrifice for their

(Source: *FM 21-76* (1957), p. 56)

country. Until this level of commitment evokes a higher standard of individual, buddy team, fire team, and squad training, their commanders should not talk too loudly about valuing human life.

# Appendix A
# Strategies for Deception

## Stratagems When in a Superior Position

*Openly Cross an Open Area*

To win consistently, one must exercise surprise. A wary foe may not fall for the usual ruses. By demonstrating in a nonthreatening way, one can cause him to relax his vigilance. When used for frequently repeated actions, he will no longer take notice of them. Then a follow-through can work. Literal translations of this axiom are, "Cross the sea without heaven's knowledge," or "Fool the emperor to cross the sea."

> [T]o cross the sea without heaven's knowledge, one had to move openly over the sea but act as if one did not intend to cross it.
> Each military maneuver has two aspects: the superficial move and the underlying purpose. By concealing both, one can take the enemy completely by surprise.... [I]f it is highly unlikely that the enemy can be kept ignorant of one's actions, one can sometimes play tricks right under its nose.[1]

> People who take ample precautions are liable to be off guard. Familiar sights do not rouse suspicion. *Yin* is the inner instead of the opposite aspect of *yang*.[2]

> One who thinks that his safeguards are well-conceived is apt to relax his vigilance. Everyday occurrences will not arouse his suspicion. Thus secret plans against him can take place in the midst of common everyday occurrences.[3]

This parable advises transferring one's own apparently relaxed mood onto the enemy. In his lessened vigilance lies opportunity.

In conflicts which involve large numbers of people, it is possible to get the opponent[s] to become lax in their guard. When they are in a state of agitation and show signs of impatience, appear as if nothing is bothering you and put forth an easygoing, relaxed stance. When you perceive that the mood has been transferred (to your opponent), you have a chance to achieve victory by making a strong attack with as much speed as possible.[4]
— Musashi Miyamoto, *Book of Five Rings*

This axiom can be applied in many ways to modern war. Security elements posing as farmers, resupply columns as refugees, and infiltrators as normal foot traffic into a city are only the most common. To keep from alienating the populace, armies will seldom treat all indigenous personnel like enemy sympathizers.

Moving about in the darkness and shadows, occupying isolated places, or hiding behind screens will only attract suspicious attention. To lower the enemy's guard, you must act in the open, hiding your true intentions under the guise of common every day activities.[5]

As offense and defense are inseparably linked, this parable applies to both. Ostensibly defending one place and occupying another is just another way of displaying a false face. Aerial photographs of WWII Axis defenses — not only Japanese,[6] but also German[7] — reveal no consistent shape.

One who is good at marshalling troops does so by putting the enemy in the unfathomable situation of fighting with shadows. He assumes no posture and reveals no shape so that there is nothing he cannot achieve. He reveals no shape and shows no move so that there is no change he cannot make. This is the supreme art of war.[8]
— Guan Zi, *Book of Master Guan*

In Vietnam, enemy soldiers sometimes ran for over 20 miles in the approach march to avoid U.S. supporting arms. On defense, the North Vietnamese liked to tunnel, just as the Japanese and North Koreans had before them. U.S. forces have unknowingly walked right past enemy fortifications in many of their wars.

A good defender hides under nine layers of earth; a good attacker moves above nine layers of heaven. Thus he is able to both preserve himself and achieve a complete victory. (Wang Xi's Note: A defender perceives no chance for attack. He conceals his shape and remains quiet so that the enemy cannot detect him. An attacker perceives a chance for attack. He moves from afar in a fabulous speed and takes the enemy by surprise so that it has not time to get ready for defense.)[9]
    — Sun Zi, *Art of War*

On a full-moon-lit night during Operation Maui Peak in Vietnam, NVA scouts dressed like bushes crawled within grenade range of U.S. troops by moving slower than perceptible to the naked eye.[10] Specially trained VC sappers could crawl unnoticed through rows of protective barbed wire.[11] Sadly, there is little official evidence of their subsequent activities. For many Americans, what has been the Asian's most powerful weapon remains an unfounded rumor.

Superb military maneuvers leave no trace. When leaving no trace, they cannot even be detected by deep-probing spies nor can be counteracted by persons of wisdom.[12]
    — Sun Zi, *Art of War*

*Seize Something Valuable the Enemy Has Left Unguarded*

It seldom works to compete head on with a more powerful opponent. While preoccupied with his next objective, he may leave behind and inadequately secure something of value. Attack it. When his forces must quit their offensive to rescue what has been lost, they will become disheartened and easy to trap. The literal translation is of this precept is, "Besiege Wei to rescue Zhao."

When the enemy is too strong to attack directly, then attack something he holds dear. Know that in all things he cannot be superior. Somewhere there is a gap in the armor, a weakness that can be attacked instead. If the enemy is on campaign, his home defense will be weak, if his army is fast, his baggage trains will be slow, if the army well equipped, the treasury will be at a loss.[13]

**243**

Again, this stratagem's application can be either physical or psychological. Its objective may be an easy-to-attack enemy convoy or an easy-to-manipulate enemy perception. One of its central themes is the application of mass. If the opposing force is too big, it must be made to disperse so that its subordinate elements can be eliminated one at a time. To get a large opposing force to spread out and expose itself, Oriental ambushers will sometimes use a few poorly aimed shots from a single rifleman. The rest of his unit can then place a large volume of fire into an open formation and sequentially eradicate isolated groups of survivors. To build momentum, the Asian knows he must string assured victories together, as opposed to impetuous moves. To win against a stronger opponent, he knows he must wait for that opponent to make a mistake.

> Instead of attacking headlong a powerful, concentrated enemy, break it up into smaller, vulnerable groups. Instead of striking first, bide your time and strike only after the enemy has struck.[14]

> It is wiser to launch an attack against the enemy forces when they are dispersed than to fight them when they are concentrated. He who strikes first fails and he who strikes late prevails.[15]

This axiom says to "relieve the besieged by besieging the base of the besiegers."[16] If one's initial skirmishes are generated in numbers greater than can be handled by an opponent's supporting-arms apparatus, then one's subsequent thrust will be unobstructed. In this way, a besieged and understrength unit can still make strategic gains. In a firefight, an Eastern force will randomly shoot from so many directions that its Western opponent cannot effectively return fire.

> Attack where he is unprepared, appear where unexpected.[17]
> — Sun Zi

*Use the Enemy against Himself*

In a military endeavor of indefinite duration, one must frugally expend his resources. Instead of endangering personnel, he must use trickery and deception to deplete the opponent's strength. One way is to encourage that opponent to hurt himself. Literally this precept translates, "Kill with a borrowed knife or sword."

> "Borrow" can mean making use of the enemy to cause its self-destruction. Use ruses to sow discord among the enemy and make them kill one another; this is making use of the enemy's own knife. To get hold of the enemy and use them is borrowing the enemy's resources. To sow discord among the enemy's generals, making them fight among themselves, is to borrow the enemy's generals. To find out the enemy's strategy and turn it to one's advantage is to borrow the enemy's strategy.[18]

While stalling, the Asian stays close enough to his foe to read intentions and capitalize on opportunity. This "close embrace" will often cause that foe to damage himself with supporting arms.

> In dialectic terms, another man's loss is your gain.[19]

> We need not act by our own [inertia] but just sit and wait for things to happen. When something is found to be difficult to do, we simply get someone else to do it.[20]
> — Bing Fa Bai Yan, *A Hundred War Maxims*

Western infantry formations often move too fast to maintain momentum. In a tropical climate, the Asian has only to wait until afternoon to confront his opponent. By then, the ill-conceived soldier's load, accelerated operational tempo, and sultry weather conditions will often have done much of his work for him.

> When you do not have the means to attack your enemy directly, then attack using the strength of another. Trick an ally into attacking him, bribe an official to turn traitor, or use the enemy's strength against him.[21]

*Make the Enemy Come to You*

While it is sometimes advantageous to hurry onto a battlefield to take advantage of terrain, weather, or some other circumstance, it is generally not a good idea.  If one's opponent already occupies that space, he better understands the situational nuances.  The idea here is to let the enemy experience the loss of energy and certainty associated with entering an unfamiliar area.  Literally the axiom translates, "Wait Leisurely for an Exhausted Enemy."

> To weaken the enemy, it is not necessary to attack him directly.  Tire him by carrying out an active defence, and in doing so, his strength will be reduced, and your side will gain the upper hand.[22]

While biding one's time, it's still important to explore — through ruses and demonstrations — the enemy's intentions and weaknesses.  This may further tire and confuse him.  There is no way a demonstration can backfire as long as its participants flee at the first indication of an enemy response.  A small force can wear out a big force by continually harassing it — alternately fighting and falling back.  A well-designed deliberate defense can withstand any number of hasty attacks from a bigger opponent until dark.  Here again is the idea of countering something with its opposite.

> Most importantly, build up one's forces before deploying them.  Use small to counter big, the unchanging to deal with the changing, the stationary to deal with the mobile.[23]

Of course, a less literal translation of this parable is also possible.  The idea here is to passively manipulate one's foe.  It will keep him tired and off balance.

> It is possible to lead the enemy into an impasse without fighting.  The active is weakened to strengthen the passive.[24]

U.S. ground forces were heavily manipulated throughout the Vietnam War.  Those attempting to use one route were lured by blatant enemy activity onto another, those nearing an enemy base camp were discouraged by booby trap mazes from entering, and those stopping for the night were kept from sleeping by intermittent mortar barrages.  When used in combination, such ruses can cause one's adversary to feel exhausted to the point of being almost powerless.

Tempt the enemy with profit to make it come forward; fore-stall it with danger to keep it from coming. In this way, we can exhaust the enemy when it is reposed, starve the en-emy when it is well fed, and provoke the enemy when it is calm.[25]

— Sun Zi, *Art of War*

*Capitalize on a Natural Disaster*

Civilizations suffer from starvation, disease, and war. The dis-ruptive influence of the first two can be exploited during the third. Literally this principle translates, "Loot a burning house."

An enemy's troubles can come from two sources. . . . Exter-nal threat. The enemy is invading our country! Internal difficulties. Natural disasters. Corrupt and dissolute offi-cials. Internal strife.[26]

When a country is beset by internal conflicts, when disease and famine ravage the population, when corruption and crime are rampant, then it will be unable to deal with an outside threat. This is the time to attack.[27]

The Western force that maneuvers during the day and stops to "medevac" casualties can be robbed of any momentum by a single sniper round or booby trap.

When the enemy falls into a severe crisis, exploit his ad-versity and attack by direct confrontation.[28]

By seeking engagement before victory, the Western commander overemphasizes body count and de-emphasizes strategy. The East-erner looks for an opponent already in disarray.

In ancient times, one who was good at warcraft attained victories that were easily attainable. Therefore, the vic-tory brought him neither fame for wisdom nor merit for courage. Each victory was certain, for it was gained by defeating an enemy that had already lost. Thus a victori-

ous army gains victory before seeking engagement whereas a doomed army seeks victory during the engagement.[29]
— Sun Zi, *Art of War*

*Pretend to Attack from One Side and Attack from Another*

The enemy will reinforce whichever side is threatened. To do so, he must reduce his vigilance in other sectors. This makes him easier to attack. Literally the phrase translates, "Make a clamor or feint to the East while attacking to the West."

> To use military strategies skillfully, move east then west, advance then retreat. Never let others know when you'll strike. When others expect you to act, don't act. When others don't expect you to act, take action. If the enemy is taken in by the stratagem and becomes confused, strike decisively to grab victory.[30]

The Asian uses ordinary forces to capture his foe's attention, and then extraordinary forces to beat him. What's visible of his initial formation is often a feint. On the Guadalcanal perimeter, the Japanese used strings of firecrackers to simulate frontal assault at one location and then sappers to penetrate another.[31] If the extraordinary forces wish to leave after accomplishing their mission, they have only to create a commotion in one sector and escape through another.

> Attack at one place and go in by another; retaliate at one place and go out by another. This is the artifice for attack and defense.[32]
> — *A Scholar's Dilettante Remarks on War*

The foe's judgment must be assessed before trying to deceive him in this way. The foe with presence of mind knows his capabilities and arrays his troops accordingly. He is less likely to be misled.

> When the enemy command is in confusion, it will be unprepared for any contingencies. . . . When the enemy loses internal control, take the chance and destroy him.[33]

(Source: Courtesy of Tony Stone Images, © Stone/Gerry Alexander)

## Stratagems for Confrontation

*Make Something out of Nothing*

"Sun Zi wrote that the direct attack and the indirect attack are interchangeable depending on the enemy's expectation."[34] Here a feint (nothing) becomes the real attack (something) and catches the foe expecting a feint by surprise. The axiom translates, "Create something from nothing."

> Make a false move, not to pass it for a genuine one but to transform it into a genuine one after the enemy has been convinced of its falsity.[35]

> Design a counterfeit front to put the enemy off-guard. When the trick works, the front is changed into something real so that the enemy will be thrown into a state of double confusion.[36]

Implicit in this precept is close attention to the opponent's disposition. An adversary who has recently occupied a defensive position will often be "trigger happy." It would therefore be unwise to attack him by whatever means the first night. Here, timing is everything.

> False and then real is the most effective. In using this stratagem, bear in mind two factors. Firstly, the character of the enemy (leader). . . . Secondly, the timing. When the enemy is blind to your intention, change from false to real and mount a surprise attack.[37]

This particular deception can be further strengthened by employing an identical feint twice in succession. The third time the opponent observes the same action, he will automatically think it a feint.

> Use the same feint twice. Having reacted to the first and often the second feint as well, the enemy will be hesitant to react to a third feint. Therefore, the third feint is the actual attack catching your enemy with his guard down.[38]

A couple of lightly armed infiltrators can often unobtrusively do what a large, well-supported formation might have trouble accomplishing with tremendous force. Before attacking *en masse* a position in Vietnam, the NVA would use one or two sappers to investigate its internal layout and a stormtrooper squad to breach its outer defenses. Often, this lone sapper or vanguard squad could destroy the position's strategic contents, thereby removing any need for the follow-up attack. In these instances, the sapper or assault squad constituted the extraordinary forces while the nondeployed backup unit was the ordinary forces. This axiom advises being able to employ extraordinary or ordinary forces on any occasion, so as to befuddle the enemy.

> One who is good at the art of war can employ either extraordinary or normal forces on every occasion, so that the enemy is deprived of judgment. Therefore he can achieve victory by either using normal or extraordinary forces.[39]
> — *Li Jing's Reply to Emperor Taizong of Tang*

*Attack from One Side and Switch to Another*

The well-seasoned opponent may not fall for a feint. One must sometimes assault him from one side and then switch to another, or at least approach him from one side and then attack on another. Literally the axiom translates, "Advance to Chencang by a hidden path."

Make a false move to tie up the main force of the enemy.[40]

To pin down the enemy, expose part of your action deliberately, so that you can make a surprise attack somewhere else.[41]

By making the enemy use normal military thinking to guess our intention, we can hoodwink them to achieve our objective. . . . Creating a scenario to feign an attack allows one to take advantage of a weak spot in the enemy's defence.[42]

This axiom meshes well with the idea of shifting the focus of main effort. It suggests frontal attack to conceal covert maneuvers to outflank the enemy. In the Orient, anything visible is often intended to deceive. For example, by avoiding contact, an Eastern vanguard might lead its pursuers into a trap. An Asian maneuver element might reconnoiter an objective along one route then attack it along another. What differentiates this stratagem from "Make a clamor in the East and attack in the West" is the idea of a converging assault.

To win victory in battle, the leader must know how to use both direct and indirect methods. The interplay between direct and indirect methods generates countless tactics.[43]
— Sun Zi

Attack the enemy with two convergent forces. The first is the direct attack, one that is obvious and for which the enemy prepares his defense. The second is the indirect attack, the attack sinister, that the enemy does not expect and which causes him to divide his forces at the last minute, leading to confusion and disaster.[44]

**251**

*Give Murphy's Law Time to Work on the Enemy*

The colloquial "Murphy's Law" states, "What can go wrong in battle will go wrong." It applies to all participants. In a confrontation, he who hesitates will often discover a weakness that his opponent did not initially possess. While the original opportunity may appear to slip away, another may arise. Frequently the goal will be revealed as not worth fighting for, or obtainable later with less effort. Literally the strategy translates, "Watch a fire from across the river."

> To remain disciplined and calm while waiting for disorder to appear amongst the enemy is the art of self-possession.[45]
> — Sun Zi

> When the discord of the enemy becomes apparent, take no action but instead wait for the oncoming upheaval.[46]

Though great significance has been attached to swiftness in military operations, a good commander should master the art of delay. Before engaging a powerful opponent, he should first discover an opportunity. Such is the basis of all delay-oriented strategies.[47] Every time a Western unit returns fire, it must work harder to maintain its alignment and thus slow down its forward progress. The resulting loss of momentum is something on which an Eastern unit can capitalize. Whenever a Western headquarters issues an order, its literal instructions will defy common sense at the squad level. For example, "Stay in your lane to attack Objective A before dark" will equate to, "Frontally attack a hidden nest of enemy machineguns while remaining fully visible yourself." The resulting friction between levels of command often leads to communication breakdowns and halfhearted execution. These are problems that can again be exploited.

> When the enemy finds itself in a predicament and wants to engage us in a decisive battle, wait; when it is advantageous for the enemy but not for us to fight, wait; when it is expedient to remain still and whoever moves first will fall into danger, wait; when two enemies are in a fight that will result in defeat or injury, wait; when the enemy forces, al-

though numerous, suffer from mistrust and tend to plot against one another, wait; when the enemy commander, though wise, is handicapped by some of his cohorts, wait.[48]
— Bing Fa Bai Yan, *A Hundred War Maxims*

This stratagem does not imply that one should sit idly by while the war goes on around him. When the time is ripe, one should strike to destroy the enemy.[49]

Delay entering the field of battle until all the other players have become exhausted fighting amongst themselves. Then go in full strength and pick up the pieces.[50]

*Mask a Sinister Intention with a Good Impression*

In war, the enemy can be lulled into a false sense of security by causing him to believe that he is in no danger. Literally the strategy advises, "Hide your dagger behind a smile."

One way or another, make the enemy trust you and thereby slacken his vigilance. Meanwhile, plot secretly, making preparations for your future action to ensure its success.[51]

When the enemy has let down his guard, hatch your scheme in secret.[52]

The stroke, when it comes, must be smooth and unobtrusive. A lot of planning — reconnaissance and rehearsal — will be required.

Reassure the enemy to make it slack, work in secret to subdue it; prepare fully before taking action to prevent the enemy from changing its mind.[53]

Throughout the Vietnam conflict, U.S. planners never suspected that enemy sapper teams and stormtrooper squads were capturing enough war materiel to partially resupply their own units and destroying enough to discourage the U.S. Congress. One truly adept at this axiom might be able to defeat an opponent while all the while leading that opponent to believe he was winning.

> Speak deferentially, listen respectfully, follow his commands, and accord with him in everything. He will never imagine you might be in conflict with him. Our treacherous measures will then be settled.[54]
>
> — *The Six Secret Teachings of Tai Gong*

Although Western forces have suffered terribly from a lack of cover in battle, they do not suffer from a lack of pride. To weaken them, their opponent has only to feign losing the preceding engagement.

> When the enemy is strong and cannot be easily overcome, we should puff it up with humble words and ample gifts and wait until it reveals its weak point to subdue it once and for all. The principle goes, "When the enemy is humble, make it proud."[55]
>
> — *A Hundred Marvelous Battle Plans*

*Sacrifice Minor Concerns for the Sake of the Overall Mission*

To gain a thing, one must sometimes lose another. Trying to hold on to everything may, in the end, cause the loss of everything. To accomplish a major goal, it may be sometimes wiser to sacrifice minor concerns. Literally the axiom translates, "Sacrifice the plum tree to save the peach."

> When the loss is inevitable, sacrifice the part for the benefit of the whole.[56]

> There are circumstances in which you must sacrifice short term objectives in order to gain the long term goal. This is the scapegoat strategy whereby someone else suffers the consequences so that the rest do not.[57]

On more than one occasion, weak forces have so tied up the opposition that other forces could attack targets of strategic significance. The idea is to risk a small unit to keep the bulk of the opposing force occupied, while gathering up other units to wipe out a lesser part of that opposing force. As long as withdrawing from a battle does not risk that battle's strategic goal, doing so can conserve limited resources. When properly covered (by terrain and fire) and prerehearsed, tactical withdrawal can be relatively safe. The Japanese occupants of each machinegun position on those Pacific islands routinely withdrew twice — the first time to escape U.S. supporting arms and the second time to avoid being overrun. They did so safely by staying at or below ground level. In Vietnam, the battlefields were not as constricted. Small enemy contingents willing to alternately fight and pull back could keep large U.S. formations at bay almost indefinitely.

> When observing the enemy's array, I ascertain its strong and weak points. Then I use my weak to meet its strong and use my strong to meet its weak. Taking advantage of my weakness, the enemy can at best advance for several dozen or a hundred paces. Taking advantage of its weakness, I always close in on it from its rear to deal a crushing blow. I have achieved most of my victories by employing this strategy.[58]
> — Li Shimin, Emperor Taizong of Tang Dynasty

> Pit your least strong horse against his strongest. Your strongest against his less strong. And your less strong against his least strong. Although I lost one race, I won two. . . . Send out our best column to quickly destroy their weakest column. Then our best can join our average column to annihilate their average column. Finally, our best and average columns can join our weakest column to wipe out their strongest column.[59]

The Asian realizes that there is no shame in strategic retreat or tactical withdrawal. He knows that both are necessary to achieve final victory against a formidable opponent with a minimal loss of life. He sees more dishonor in refusing to withdraw when it is strategically wise to do so.

> Know contentment, and you will suffer no disgrace. Know when to stop, and you will meet with no danger. You can then endure.[60]
> — Lao Zi

One who looks into the distance overlooks what is nearby; one who considers great things neglects the details. If we feel proud of a small victory and regret over a small defeat, we will encumber ourselves and lose the opportunity to achieve merits.[61]
— Li Su, a Tang Dynasty general

*Seize the Chance to Increase the Odds*

Victory is not totally assured to the most diligent of military planners. Chance always enters into the equation, so the window of opportunity must be sought out and quickly exploited. Literally this axiom means, "Seize the opportunity in passing to lead away the sheep."

Take advantage of the smallest flaw; seize the smallest profit. Make use of a minor mistake of the enemy to gain a minor victory.[62]

Exploit any minor lapses on the enemy side, and seize every advantage to your side. Any negligence of the enemy must be turned into a benefit for you.[63]

Again implicit in this axiom is the need for timely action. This strategy requires one to be totally responsive to changing circumstances.[64]

Act quickly when perceiving the advantage; halt when there is no advantage. To pursue an advantage and take an opportunity does not allow the error of a moment of breath. It is too early to act an instant earlier and too late to act an instant later.[65]
— *Summary of Military Canons*

The odds of success for any attack depend heavily on its degree of surprise. Oriental maneuver elements (unlike their Western counterparts) will sometimes abort an ongoing attack. If they cannot maintain or regenerate enough surprise, they will temporarily withdraw and resume the attack from another direction later.

To seize opportunity, nothing is more significant than the element of surprise. Therefore, when losing its natural defenses, a fierce beast may be driven away by a child with a spear. . . . This is because the danger emerges too suddenly for one to retain one's presence of mind.[66]
— Jiang Yan, *The Art of Generalship*

Take action when an advantage is perceived and halt when there is no advantage.[67]
— Tai Bai Yin Jing, *The Yin Canon of Vesper*

Oriental armies have learned to exploit the slow and methodical way in which U.S. forces attack without much reconnaissance or rehearsal.

Just before a large army enters battle, its shortcomings will be exposed. Exploiting them will lead to victory.[68]

Taking a goat in passing without being noticed is tantamount to striking a blow in passing without being felt. If this were possible, the same subtle raid could be used on the same objective over and over. Allied Headquarters in Vietnam never realized that U.S. bases were being repeatedly overrun by tiny units. Base incursions produced few bodies, American or otherwise, and were therefore assumed to be unsuccessful probes. The squads overrunning those bases were not interested in killing defenders, only in destroying targets of strategic significance. They would "secretly penetrate by force" one side of a perimeter (using an indirect-fire deception), blow up a command bunker or artillery piece, throw down an 82mm mortar fin, and then sneak out the other side. When, in the morning, U.S. defenders could find only a couple of enemy bodies in the wire and evidence of a lucky mortar hit, they would naturally assume that they had beaten off another attack.

While following the rules of strategy and tactics, be prepared to take advantage of circumstances not covered by conventional thinking. If opportunities present themselves, then the leader should be flexible in his plans and adapt to the new circumstances.[69]
— Sun Zi

(Source: Courtesy of China Pictorial ©)

## Stratagems for Attack

*Make a Feint to Discover the Foe's Intentions*

The inexperienced or nervous opponent may overreact to a feint and thereby reveal his overall plan. In war, thwarting an enemy's strategic intentions can be as beneficial as accomplishing one's own. In Chinese, the axiom reads, "Beat the grass to startle the snake."

> When you cannot detect the opponent's plans, launch a direct, but brief, attack and observe your opponent's reactions. His behavior will reveal his strategy.[70]

The Asian will routinely use a sucker move to discover Western intentions. He attempts not always to act first, but always to make the first successful follow-through. To do so, he must discover what his opponent has in mind.

> In military terms, this strategy refers to finding out about the enemy intentions before making a move.[71]

> Any suspicion about the enemy's circumstances must be investigated. Before any military action, be sure to ascertain the enemy's situation; repeated reconnaissance is an effective way to discover the hidden enemy.[72]

There are any number of ways to apply this axiom in war. One of the more common is deploying a large force to attract attention and a small force to secretly reconnoiter or do damage. Sometimes the enemy never realizes that the small force even exists.

> In warfare, employ both large and small forces to observe the enemy's reaction, push forward and pull back to observe the enemy's steadiness, make threats to observe the enemy's fear, stay immobile to observe the enemy's laxity, take a move to observe the enemy's doubt, launch an attack to observe the enemy's solidity. Strike the enemy when it hesitates, attack the enemy when it is unprepared, subdue the enemy when it exposes its weakness, and rout the enemy when it reconnoiters. Take advantage of the enemy's rashness, forestall its intention, confound its battle plan, and make use of its fear.[73]
> — Sima Fa, *Law of Master Sima*

Sometimes, the Asian will use an entire battle to discover the fighting capacity of his opponent. It has been said that the battle of Ia Drang may have been like that in Vietnam. The side that occupies the ground after a battle is only the victor if that ground happens to have strategic value.

> If we and the enemy have not engaged yet and are therefore unacquainted with each other, we can sometimes dispatch a band of troops to test the enemy's strength and weakness; this is called a tasting battle. For the tasting

battle, we should allow it to be neither grave nor long and should withdraw the troops after a brief engagement. Coordinating forces should be sent to cope with emergencies and prevent the loss of the tasting army, which may bring about the defeat of our main force.[74]
　　　　　— Wu Bei Ji Yao, *An Abstract of Military Works*


*Steal the Enemy's Source of Strength*

Philosophies, institutions, traditions, and procedures all have varying degrees of moral and emotional power. Those of the enemy can be appropriated. For example, Mao Tse-tung — an avowed atheist — treated prisoners well in his war with Chiang Kai-shek to stimulate recruitment and dispel rumors of barbarism.[75] This principle can be translated, "Find reincarnation in another's corpse," or "Raise a corpse from the dead."

> Take an institution, a technology, or a method that has been forgotten or discarded and appropriate it for your own purpose. Revive something from the past by giving it a new purpose or to reinterpret and bring to life old ideas, customs, and traditions.[76]

Embedded in this axiom is the idea of looking for little ways to manipulate one's opponent and thereby keep him off balance. With relatively more individual and small-unit skill, even a severely depleted Oriental unit can contribute to the overall war effort.

> The powerful is beyond exploitation, but the weak needs help.[77]

> Instead of being controlled by others, be in control. . . . In military strategy, it means using all available means to boost one's standing.[78]

To successfully apply this parable, one must concentrate on the difference between truth and fiction. Because the Eastern leader readily acknowledges when his unit has been beaten, he does more to correct its deficiencies.

In warfare there are no ever-victorious generals but there
are many who taste defeat.[79]

By focusing on what the individual soldier and small unit can
accomplish in the face of massed firepower, the Asian creates the
impression of righteousness among his troops and the war zone
population alike. He knows that victory will depend on winning
the hearts and minds of the people.

> The pivot of war is nothing but name and righteousness.
> Secure a good name for yourself and give the enemy a bad
> name; proclaim your righteousness and reveal the
> unrighteousness of the enemy. Then your army can set
> forth in a great momentum.[80]
> — *A Scholar's Dilettante Remarks on War*

## Draw the Enemy Away from His Refuge

As a rule, one does not attack an opponent who has the edge in
terrain, defensive works, or home ground. Through trickery and
deception, that opponent must be lured away from his comfortable
sanctuary. In Chinese, this axiom says, "Lure the tiger off the mountain."

> With regard to heights, if you occupy them before the en-
> emy you can wait for the enemy to climb up. But if he has
> occupied them before you, do not follow but retreat and try
> to entice him out.[81]
> — Sun Zi

> Never directly attack a well-entrenched opponent. Instead
> lure him away from his stronghold and separate him from
> his source of strength.[82]

Implicit in this stratagem is the kind of ground suitable for the
trap. Oriental ambushes often start out as sniping incidents to get
American forces to fully deploy across relatively open ground. At
that point, the U.S. forces become vulnerable to grazing machinegun
fire from the flanks.

> Use unfavorable natural conditions to trap the enemy in a difficult position. Use deception to lure him out. In an offensive that involves great risk lure the enemy to come out against you.[83]

> Use terrain to trap the enemy. Use deception to lure him. The "tiger" refers to a strong enemy. The "mountain" refers to his bastion. Lure the enemy out to fight in a place where you hold the advantage. The tiger loses on level ground.[84]

Also implicit in this axiom is the advantage gained from making an opponent move onto unfamiliar ground. The enemy is lured away from his impregnable position. In Vietnam, U.S. forces got beaten routinely during search and destroy missions. This fact never surfaced because, after winning, the foe would relinquish ground of no strategic value.

> If the enemy is led on to the battlefield, its position is weak. If we do not have to reach for the battlefield, our position is always strong. Use various methods to make the enemy come forward and lie in wait for it at a convenient locality; we can thereby achieve certain victory.[85]
> — *A Hundred Marvelous Battle Plans*

*Give a Retreating Adversary Room*

A desperate fugitive may turn to fight. By letting him go, one can reduce his eagerness for a last-ditch stand. Literally the axiom reads, "To catch something, first let it go."

> Cornered prey will often mount a final desperate attack. To prevent this you let the enemy believe he still has a chance for freedom. His will to fight is thus dampened by his desire to escape. When in the end the freedom is proven a falsehood the enemy's morale will be defeated and he will surrender without a fight.[86]

Inherent to this principle is patience. The enemy's will to fight is an important objective.

Do not obstruct an army retreating homeward. If you besiege an army you must leave an outlet. Do not press an
exhausted invader.[87]
— Sun Zi

Press enemy forces too hard and they will strike back
fiercely. Let them go and their morale will sink. Follow
them closely, but do not push them too hard. . . . In short,
careful delay in attack will help to bring destruction to the
enemy.[88]

In warfare, if the enemy is outnumbered by our troops, it
will be afraid of our strength and flee without fighting. We
should not embark in hot pursuit, for anything forced to
the extreme will develop into its opposite.[89]
— *A Hundred Marvelous Battle Plans*

This precept can be interpreted in many ways. By destroying
targets of strategic significance without killing defenders, Asians
have lessened the resolve with which materiel-rich Westerners defend those targets.

By making the enemy feel lucky to get away, he'll lose the
will to put up a good fight.[90]

*Discover a Foe's Intentions by Offering Him Something of Value*

He who knows where his foe plans to move next has the edge.
To find out, he can offer the opponent an incentive to go there.
Literally the axiom translates, "Cast a brick to attract a gem."

One needs to bait the fish, so hook the enemy with the prospect of gain.[91]

Lure the enemy with counterfeits.[92]

Implicit in this axiom is the idea of manipulating the whereabouts of one's opponent. Great advantage lies in choosing where
a battle will be fought. Detailed knowledge of the terrain can lead
to every square inch of ground being covered by fire.

[A]gainst the invasion of a hostile power, one should refrain from launching an attack into the enemy territory.
Instead, it is best to allow the invading army to move in
and then defeat it on one's own ground. . . . First, the commander offers the enemy some bait, which can be a body of
weak troops, [or] poorly guarded provision(s). . . . At the
prospect of gain, the enemy will advance to swallow the
bait. Thus the commander has gained the initiative by
maneuvering the enemy at his will, and the battle has actually been half won before it is fought.[93]

Which incentive to offer depends on what the opponent deems
important. The chance to corner a large number of enemy soldiers
has more than once been enough to draw an American unit onto
unfamiliar ground.

Abandon goods to throw the enemy into disorder, abandon
troops to entice it, and abandon fortresses and land to encourage its arrogance.[94]
— Bing Fa Bai Yan, *A Hundred War Maxims*

The Asian will sometimes bait his trap with the prospect of
capturing two or three fleeing soldiers. Whether or not the Western unit pursues them, it will still be mystified by their behavior.

Use bait to lure the enemy and take him in.[95]

Prepare a trap, then lure your enemy into the trap by using bait. In war the bait is the illusion of an opportunity
for gain.[96]

Bait them with the prospect of gain, bewilder and mystify
them.[97]
— Sun Zi

The Asian stays hidden and refuses to defend unimportant terrain; he thereby ensures surprise for those engagements of strategic value.

One who is good at maneuvering the enemy makes a move
so that the enemy must make a corresponding move, offers

bait so that the enemy must swallow it, or lures the enemy with prospect of gain and waits for it with one's main force.[98]
— Sun Zi, *Art of War*

## *Damage the Enemy's Method of Control*

An enemy unit's effectiveness can be lowered by disrupting its command-and-control apparatus. Destroying its command group will leave the unit leaderless. Literally the axiom translates, "To catch rebels, first catch the ringleader."

Destroy the enemy crack forces and capture their chief, and the enemy will collapse.[99]

A good commander must grasp the overall situation and strike a decisive blow to the enemy.[100]

The Asian will try hard to kill the commander of a centrally controlled, Western unit. On defense, he will stay hidden until that commander comes within view or dispatch a roving sniper. A unit's command-and-control apparatus has strategic significance.

When our troops have penetrated deep into the enemy's territory and the enemy strengthens its defense works and refuses to engage in battle for the purpose of wearing us down, we may attack its sovereign, storm its headquarters, block its return route, and cut off its provisions. Thus it will be compelled to fight, and we can employ crack forces to defeat it.[101]
— *A Hundred Marvelous Battle Plans*

There are ten thousand artifices of war, and one should not stick to any one of them. First seize the enemy's mainstay, and its strength will be weakened by half.[102]
— Hu Qian Jing, *Canon of the General*

The Eastern commander tries first to snipe, then to confuse, and finally to demoralize his Western counterpart. He accomplishes the latter by appearing to win every fight and to take fewer losses. He has learned how to undermine the pride of an attritionist foe.

When asked, "If the enemy troops are superior in number and about to advance in an orderly formation, how shall I cope with them?" I reply, "Seize something they treasure and they will become maneuverable."[103]
— Sun Zi, *Art of War*

But killing the opposing commander does not always work to one's advantage. His death may cause his troops to fight harder.

If the enemy's army is strong but is allied to the commander only by money or threats then, take aim at the leader. If the commander falls, the rest of the army will disperse or come over to your side. If, however, they are allied to the leader through loyalty then beware, the army can continue to fight on after his death out of vengeance.[104]

(Source: Courtesy of Stefan H. Verstappen, © 1999)

## Stratagems for Confused Situations

*Erode the Enemy's Source of Strength*

An opponent's strength often lies in his wealth, resources, or manpower. If in wealth, cause him to incur expenses; if in resources, disrupt the lines of distribution; if in manpower, sow discord. Literally the stratagem proposes, "Remove the firewood from under the cauldron."

> Avoid a contest of strength with the enemy but seek to weaken his position.[105]

> When confronted with a powerful enemy, do not fight them head-on but try to find their weakest spot to initiate the collapse.[106]

This axiom can be applied quite literally. Concerted reconnaissance may locate relatively exposed items of strategic importance. Stealing or destroying them will greatly weaken a more powerful opponent.

> In large scale battles, after careful inspection of the enemy's forces, one can gain advantage by attacking the corners of exposed strategic points. When one has eliminated the strength of the corners, the strength of the whole will also be diminished.[107]
> — Musashi Miyamoto, *Book of Five Rings*

The axiom can also be applied psychologically. Oriental defenders can sap the morale of Western attackers by simply moving into a secret chamber below ground about to be captured. With too few bodies to show for the effort, the attackers feel beaten. By fighting just long enough to inflict casualties, the Easterner tempts his Western counterpart to exaggerate the final body count.

> Don't fight a powerful enemy head-on, instead undermine its morale and deprive it of leadership. . . . Study the war situation carefully to find out the enemy's weak points and exploit them.[108]

When faced with an enemy too powerful to engage directly you must first weaken him by undermining his foundation and attacking his source of power.[109]

This part of the Asian way of fighting is perhaps the most difficult to fathom. In essence, the Oriental frontline commander attempts — with what he does on the battlefield — to create for his Western opponent a moral dilemma. It must be part of the overall strategy on which he has been briefed.

In directing warfare and assessing the enemy, one should try to undermine the enemy's morale and destroy its discipline, so that it looks intact but loses its utility. This is the method to win by political strategy.[110]
— Wei Liao Zi, *Book of Master Wei Liao*

*Create Chaos to Make an Opponent Easier to Beat*

People will watch anything unusual in their environment. Magicians, card sharks, pick pockets, and prize fighters rely on this trait. They can redirect their quarry's attention while secretly achieving their agenda. Literally the stratagem translates, "Muddle the water to seize fish," or "Fish in troubled waters."

Before engaging your enemy's forces, create confusion to weaken his perception and judgment. Do something unusual, strange, and unexpected as this will arouse the enemy's suspicion and disrupt his thinking. A distracted enemy is thus more vulnerable.[111]

When the enemy falls into internal chaos, exploit his weakened position and lack of direction and win him over to your side.[112]

Whether on ambush, defense, or attack, Asians will often confuse their quarry by alternately firing from different locations. That quarry has difficulty locating the most damaging fire.

Every day have the vanguard go forth and instigate skirmishes with them in order to psychologically wear them

out. Have our older and weaker soldiers drag brushwood to stir up the dust, beat the drums and shout, and move back and forth. . . . Their general will certainly become fatigued, and their troops will become fearful. In this situation the enemy will not dare to come forward. Then when we come forth with our three armies the enemy will certainly be defeated.[113]
— *The Six Secret Teachings of the Tai Gong*

Much of what happened in Vietnam may have been meaningless posturing. Meanwhile, enemy maneuver elements secretly penetrated U.S. perimeters, secretly destroyed strategic U.S. targets, and then secretly left without killing U.S. soldiers.

One who is good at combatting the enemy fools it with inscrutable moves, confuses it with false intelligence, makes it relax by concealing one's strength, causes it to hesitate by exposing one's weakness, deafens its ears by jumbling one's orders and signals, blinds its eyes by converting one's banners and insignias, eases off its vigilance by hiding what it fears, saps its will by offering what it likes, confounds its battle plan by providing distorted fact, and breaks its courage by showing off one's power.[114]
— *A Scholar's Dilettante Remarks on War*

Create chaos in the enemy camp and when it's weak and without direction, you can hold sway over them.[115]

## Leave Behind a Small Force to Slow and Deceive Pursuers

Without enough discipline, retreating troops are highly vulnerable. A rear guard can provide valuable security. It can also create the impression that no retreat is underway. Literally the axiom translates, "The cicada sheds its skin."

When you are in danger of being defeated, and your only chance is to escape and regroup, then create an illusion. While the enemy's attention is focused on this artifice, secretly remove your men leaving behind only the facade of your presence.[116]

It is important for the rear guard to closely mimic the front array of the departing unit. To do so, it may construct human dummies. Or it may just have each member man more than one defensive position.

> Make your front array appear as if you are still holding your position so that the allied forces will not suspect your intention and the enemy troops will not dare to attack rashly. Then withdraw your main forces secretly.[117]

Inherent to this axiom is the idea of prompt but orderly withdrawal. The false front will do little good if the enemy sees the rest of the unit falling back. One must pull away quickly while maintaining the appearance of inaction.[118]

> To "shed the skin" doesn't mean doing so out of panic, but keeping up appearances to preserve oneself while retreating, so that the enemy won't suspect anything amiss.[119]

The false front must simulate more than just the visible parts of the original formation; it must also imitate its ongoing actions. When encircled, an Oriental force will sometimes withdraw after directing machinegun fire at the opposing-force commander. It does so to keep its movement from being observed by someone who could quickly summon supporting arms. In early 1969, a U.S. Marine battalion-sized sweep trapped an NVA unit in the village of "Chau Phong (1)" northwest of An Hoa. Reading too much into the automatic-weapons fire coming from the occupied hamlet, the Americans assaulted to find no one there. The enemy's main force had probably escaped through a gully before the U.S. cordon could tighten. A lone machinegunner may have kept the Americans' heads down and then hidden below ground.[120]

> Maintain the original shape and play out the original pose, so that . . . the enemy does not move.[121]

*Encircle the Foe but Let Him Think He Has a Way Out*

While dangerous to corner an opponent, it is not without risk to let him escape. Leaving him a way out of his predicament and then removing that way will erode his will to fight. In Chinese, the axiom says, "Bolt the door to catch the thief."

> When dealing with a small and weak enemy, surround and destroy him. If you let him retreat, you will be at a disadvantage pursuing him.[122]

> If the enemy is weak, surround them and finish them off. If the enemy is small in number, it is not wise to pursue for they're nimble in movement.[123]

U.S. patrols and maneuver elements routinely consider secure the way they enter a location. To them, their approach route constitutes a viable avenue of egress. However, a slightly larger enemy force can block their escape psychologically as well as physically. Not knowing where the fatal bullet is coming from accomplishes the former. In Vietnam, enemy maneuver elements routinely "double-enveloped" any U.S. patrol they encountered. Once encircled, the American troops faced a gradually tightening noose of crawling opponents who never shot from the same place twice. Many failed to survive the experience.

> When one fights an opponent and it appears on the surface that he has been defeated, if his fighting spirit has not yet been eradicated in his heart of hearts, he will not acknowledge defeat. In that case, you must change your mental attitude and break the opponent's fighting spirit. You must make him acknowledge defeat from the bottom of his heart. It is essential to make sure of that.[124]
> — Musashi Miyamoto, *Book of Five Rings*

Encirclement plays a significant role in the Oriental planned ambush and deliberate defense. Sometimes the quarry is allowed to move right over the top of well-hidden emplacements. Then the back door is slammed shut. Of course, the escape route does not have to be physically blocked. It can be covered by long-range fire or lead to indefensible terrain or to an untenable circumstance.

The strategy points out that encirclement and annihilation is *[sic]* the best method to deal with such a guerrilla force. . . . For instance, on learning of the enemy's plan for a night attack on your camp, you can pull away the troops and leave the camp empty. When the enemy enters the camp, you can lead the troops to close in from all sides and form a tight encirclement. Or you can use a bait army to lure the enemy into an ambush ring that allows no way of escape.[125]

## *Concentrate on the Nearest Opposition*

How well a foe fares in battle depends on how quickly he can concentrate and disperse his forces. To beat him, one must develop momentum through back-to-back victories. Taking a quick local victory, however small, can be the first step. With initial momentum, more difficult, distant targets become attainable. Here, final intentions are both facilitated and masked by exploiting a local opportunity. Literally translated the axiom reads, "Befriend a distant state while attacking a nearby state."

It is more advantageous to conquer the nearby enemies, because of geographical reasons, than those far away.[126]

Obviously, additional advantage lies in preempting the intentions of one's nearest opponent. A surprise attack from him could jeopardize one's overall plan.

When you are the strongest in one field, your greatest threat is from the second strongest in your field, not the strongest from another field.[127]

In strategy, totally different approaches can be used to achieve one and the same purpose.[128]

Stressed is the need to beat a powerful opponent piecemeal. Avoided are battles that don't contribute to the overall war effort.

[T]he strategy instructs the military leader to deal with his enemies one by one. Also, one is cautioned against seeking superficial victories that do not bring about any concrete profit.[129]

Of course, this axiom can also be applied in a less literal context. Sapper teams and stormtrooper squads depend for much of their success on microterrain. Western forces could usefully pay more attention to the tactical value of their surroundings. Moving one's attack point or defensive position 100 yards to the left or right can sometimes make all the difference. Headquarters seldom restricts movement to that extent. He who cannot recognize nearby opportunities may lack the flexibility to take minimal casualties in war.

> When circumscribed [submersed] in situation and restricted in disposition [posture], seek profit from nearby and keep peril at a distance.[130]

### Borrow from the Enemy the Instrument of His Own Destruction

In war, there are two sources of opposition — the human adversary and unfavorable circumstance. Here one reduces his own shortfall by appropriating part of an enemy's strength and using it against him. In this way, additional assets are generated. One can take something from his adversary to reduce a shortage of materiel, for example, and then kill him with it. In Chinese, the precept reads, "Borrow a route to conquer Guo."

> Borrow the resources of an ally to attack a common enemy. Once the enemy is defeated, use those resources to turn on the ally that lent them in the first place.[131]

Literally the principle says to take from one's more wealthy opponent whatever is needed to win the war. Rumored to have been captured early in the Vietnam conflict was an Allied convoy carrying enough claymore mines to resupply the enemy for the rest of the war.[132] With the Allies providing a continual source of resupply, why would Ho Chi Minh hesitate to send reinforcements South?

> When a small state, located between two big states, is being threatened by the enemy state, you should immediately send troops to rescue it, thereby expanding your sphere of influence.[133]

When somebody wants to use your territory to deal with an external threat, do not believe him so readily.[134]

Of course, the axiom can also be applied more subtly, like borrowing an opponent's appearance. Presumably to generate more indecision and less fire, NVA soldiers dressed like U.S. Marines on not only Operation Buffalo,[135] but also Mameluke Thrust.[136] Oriental infiltrators have used the same ploy.

Good intentions can be borrowed as well. When surrounded, an Oriental unit has more options than one might imagine. First, small groups monitor the cordon's location and exfiltrate U.S. lines. Then, what's left of the unit hides in spider holes (or below ground) while a few volunteers with automatic weapons create the illusion of a last stand somewhere else. The cordon tightens around the mock machinegunners — leaving the spider hole occupants behind U.S. lines. If, then, the rear-guard volunteers can hold out until dark, everyone escapes. The volunteers have made their "last stand" at the easily hidden mouth to a subterranean room or tunnel.

(Source: Corel Print House, *People*, Sit_Sam)

## Stratagems for Gaining Ground

*Attack the Enemy's Habits to Undermine His Foundation*

In combat, a large unit's actions are determined more by terrain and habit than by what's logical. A unit invincible in one formation may be totally ineffective in another. By altering the rules and habits under which it fights, one can take away its physical and moral foundation. Literally the axiom translates, "Steal the beams and replace the pillars with rotten timber."

> When confronting a powerful enemy, adopt feints, sudden maneuvers, and split the enemy strength to weaken it.[137]

> Disrupt the enemy's formations, interfere with their methods of operations, change the rules in which they are used to following, go contrary to their standard training. In this way you remove the supporting pillar, the common link which makes a group of men an effective fighting force.[138]

While this ploy could be applied to any of an opponent's rules or habits, it appears to be aimed primarily at his formation. Of primary interest is causing that formation to frequently change and subsequently fragment. The goal seems to be separating the main body from its march security elements. Without advance and rear-guard protection on a straight trail, for example, a main body could be mowed down with one burst from a well-positioned machinegun.

> Change the enemy's formation frequently, dislocate its main force and deal the blow when it tends toward defeat.[139]

> A typical battle formation has a central axle (heavenly beam) extending from the front to the rear and a horizontal axle (earthly pillar) connecting the right and left flanks. The two axles are made up of the best fighters.[140]

Easterners sometimes use a series of sudden maneuvers to thoroughly confuse a foe. On ambush, they may only start with a feigned sniper attack from the front. Then, a full-blown assault may drive the quarry back into sniper fire from spider holes at its rear.

Make the allied forces change their battle formation frequently so that their main strength will be taken away.[141]

The axiom further advises the skilled warrior in an unfavorable position to proceed slowly.[142] By hiding in a spider hole, an Asian can bypass the security elements of a U.S. maneuver element to get at its armor, main body, or command group. Trying to slow him down with a Western-style defense has been likened to trying to block water with a fish net.

*Make an Example of Someone to Deter the Enemy*

A bigger foe will sometimes advance too cautiously after seeing the destruction of one of its subordinate elements. In Chinese, this axiom reads, "Point at the mulberry tree and abuse the locust."

To discipline, control, or warn others whose status or position excludes them from direct confrontation, use analogy and innuendo.[143]

Feigning an attack from the front will only momentarily distract the foe's attention away from his other sides. For this reason, the Asian will occasionally launch a frontal, human-wave assault. By so doing, he causes his future adversary to worry less about flanks and rear.

To dangle a carrot may not be enough to win someone over, for he may be suspicious of your motives. But if you warn him subtly by using another as an example, he may take the hint.[144]

One's uncompromising stand will often win loyalty, and one's resolute action, respect.[145]

*Feign Lack of Military Ability*

Irrational behavior normally generates a sound or motion signature. But one can unobtrusively feign tactical ignorance. Literally this axiom says, "Feign foolishness instead of madness."

At times, it is better to pretend to be foolish and do nothing than brag about yourself and act recklessly. Be composed and plot secretly, like thunder clouds hiding themselves during winter only to bolt out when the time is right.[146]

In Vietnam, the enemy threw satchel charges into U.S. perimeters that contained no shrapnel. At the time, the American defenders thought their foe to be stupid. As it turns out, the satchel charges were not intended to hurt anyone, only to keep their heads down during an assault. They were concussion grenades large enough to imitate impacting artillery shells.

Secrecy begets success; openness begets failure. In military conflicts, it's better to conceal than to reveal moves, to play dumb than to act smart.[147]

In modern war, the side that first reveals its shape and intentions is often the side that loses. In every ostensible act, fake or otherwise, lies some clue to one's overall plan. Total secrecy is frequently preferable to some type of deception. Of utmost importance, from the standpoint of an Asian, is remaining motionless upon the enemy's arrival.

Stay motionless and hide one's intention.[148]

To undertake military operations, the army must prefer stillness to movement. It reveals no shape when still but exposes its shape in movement. When the enemy exposes a vulnerable shape, seize the chance to subdue it.[149]
— Bing Lei, *Essentials of War*

In Vietnam, enemy forces often appeared unwilling to confront their U.S. counterpart. They would let the U.S. forces come to them and then watch for an opening.

Hide behind the mask of a fool, a drunk, or a madman to create confusion about your intentions and motivations. Lure your opponent into underestimating your ability until, overconfident, he drops his guard. Then you may attack.[150]

*Lure the Foe into Poor Terrain and Cut Off his Escape Route*

Terrain can offer opposite advantages to different-sized forces. While narrow terrain will restrict a large force, it can protect a small force. The idea here is lure the enemy into terrain that limits his activities while helping one's own. It translates, "Climb onto the roof and remove the ladder."

> With baits and deceptions, lure your enemy into treacherous terrain. Then cut off his lines of communication and avenue of escape. To save himself he must fight both your own forces and the elements of nature.[151]

Of course, there are many enticements possible. In the Orient, the most common may be appearing to have lost the last engagement and thereby leading one's opponent to believe he can win. If he immediately exploits his "success," he may get overextended. It is then that his life line can be severed.

> Ladder refers to the means by which the enemy is lured into the ambush ring. It may be a gesture of weakness if the enemy commander is arrogant, an enticement of gain if the enemy commander is greedy, or repeated defeats if the enemy commander is hesitant.[152]

> Provide the army with an apparent chance, entice it to advance, cut it off from coordination or reinforcement, and place it in an impasse.[153]

In one interpretation, the foe's retreat can be impeded by the terrain itself. Of interest are the kinds of terrain with this potential.

> Avoid terrain that features cliffs and crags, narrow passes, tangled bush, and quagmires. While avoiding such places ourselves, try to lure the enemy into such areas so that when we attack the enemy will have this type of terrain at his back.[154]
> — Sun Zi

Normal terrain that has been covered by fire — a "firesack" —

will also work. It may be what appears to be a gap in one's own lines. Then, intruders are quickly decimated by grazing machinegun fire from both flanks. In the matrix of small forts that replaced defensive lines in WWI, every opening between forts constituted a firesack. WWII Americans often successfully assaulted dummy lines only to find themselves caught in the crossfire between enemy squad- or platoon-sized perimeters. Sometimes, they were allowed to advance past the first row of heavily camouflaged forts unchallenged. Then, when the enemy appeared at close range from all sides, the Americans were effectively cut off from reinforcement or resupply.

> Expose your weak points deliberately to entice the enemy
> to penetrate into your line, then ensnare him in a death
> trap by cutting off his rear-guard support.[155]

The Asian will routinely place a row of forts on the reverse slope of a defended ridgeline. Carefully hidden, those forts will be easy to overlook. After fighting up the front face of a contested hill or ridgeline, Western soldiers might not expect its rear slope to be still occupied. When they tangle with the next forward-slope defense, they may discover fire coming from all directions. Often, the Asian's forward- and reverse-slope positions are connected by tunnel. This strategy can also be used to instill more fighting spirit in one's own forces.

> In battle, the "ladder" refers to deliberately exposing a weak
> spot to entice the enemy to advance towards you. Once he
> has entered your ambush ring, cut off his escape route. . . .
> The strategy can be used to entice and destroy the enemy,
> and to put one's own soldiers in a position with no way of
> retreat so that they'll fight with all their might.[156]

*Mislead the Enemy with False Information*

While it is important to discover the foe's intentions, it is also important to disguise one's own. Through the use of decoys, facades, and camouflage, one can conceal his own strengths and weaknesses. Literally the axiom reads, "Tie silk blossoms to a dead tree."

First, from nature, there are a myriad of ways to disguise intentions. Screening one's activities with heavy vegetation is only the most obvious. The Easterner has taken camouflage to a new plateau. While crouching, his soldiers are almost impossible to see in most types of terrain. On Operation Buffalo in Vietnam, they managed to look like elephant grass during fire and movement.[157] The opponent is allowed to see only what might lead him to misjudge the shape of the Oriental formation. Western forces often think they are facing a defensive line, when really about to enter a lightly occupied area swept by long-range fire from either flank.

Use deceptive appearances to make your troop formation look much more powerful than it really is.[158]

Then there are the man-made noises that can mask the sounds of battlefield movement. To convince Western defenders to stay put while their relief column gets ambushed, an Asian might employ tracked vehicle noises during an intense artillery barrage. At least this is what happened to 1st Battalion, 4th Marines at Gio Linh in the early spring of 1967.[159]

Deceptive appearances coupled with unusual noises can make up for the lack of military might to subdue the enemy.[160]

There are, of course, an infinite number of other ways to look stronger than one really is.

Tying silk blossoms on a dead tree gives the illusion that the tree is healthy. Through the use of artifice and disguise make something of no value appear valuable; of no threat appear dangerous; of no use, useful.[161]

Flaunting one's strength will capture the enemy's attention and create opportunities for extraordinary forces.

Displaying power to intimidate the enemy is a timeless strategy. . . . Feign advances that you have not secured, achievements that you have not made, strength that you lack, and maneuvers to deceive the enemy. Flaunt your

strength, subdue the spirit of the enemy, and use extraordinary plans to secure victory.[162]
— Bing Fa Bai Yan, *A Hundred War Maxims*

Appearing to be exceedingly strong can also lower enemy morale. Infiltrators raising their flag at the center of a city might cause the defenders at its periphery to lose heart. The NVA flag flew over Hue City's imperial palace for three weeks in February of 1968. It was eventually taken down, but not before the illusion that America was winning the war had been dispelled.

Spreading out pennants and making the flags conspicuous are the means by which to cause doubt in the enemy. Analytically positioning the fences and screens is the means by which to bedazzle and make the enemy doubtful.[163]
— Sun Bin

There are a number of man-made devices with which to perplex an enemy — everything from prefabricated dummies to camouflaged suits. The list is practically limitless. The Viet Cong routinely built bamboo bridges a foot or two below the waterline of rivers.

Victory in war can be achieved by [one's] ingenuity. One who is conversant with a minor skill can launch great causes, and one who is versed in a minor art can attain great achievements. No skill is too lowly and no art too humble. . . .
. . . [B]ind artemisia into human figures, set up fences into a city, and fasten grass into a battle formation: We can thereby perplex the enemy by false shapes.
Direct river courses, tunnel through great mountains, move away small mounds, and transfer bridges: We can therefore ward off harm. . . .
. . . [S]et the slings on fire before hurling them: We can thereby enhance our capabilities.
Tie up chickens to start fire, lock up pigeons to reconnoiter, train monkeys to surprise the enemy's camp, drive beasts to break out of an encirclement: We can thereby exploit the use of various things.
Splash water to freeze the ground, spread dust to con-

ceal the marshland, build a deceptive bridge to trap the enemy cavalry, and dig gullies to enmesh the enemy troops: We can thereby hurt the enemy by various traps.

Set the enemy's ships on fire, conduct water to flood its city . . . , toss hooks to seize objects: We can thereby compensate for the inadequacy of our weapons.

. . . [A]ccumulate ice into a rampart. . . . We can thereby supplement the incapacity of our weapons.

. . . [T]hings can be transformed and moved, and people can appear and vanish: We can therefore fool the enemy and deprive it of judgment.[164]

— *A Scholar's Dilettante Remarks on War*

Of course, the most effective types of misinformation are those that are strikingly unique. Acting out — for everyone to see — a mistake that occurs only rarely in one's own army could cause an opponent to badly underestimate that army's level of discipline. Throughout most of the Vietnam War, U.S. troops believed their Eastern counterparts to be running away because they lacked the courage, proficiency, and wherewithal to confront U.S. forces. They had been conditioned to think that professional warriors never move backwards.

People are accustomed to what they often see and are startled by what they rarely see.[165]

— Bing Lei, *Essentials of War*

One wonders how many of the accidents of war — like the "short rounds" that plagued U.S. forces in Vietnam — may have been indirectly caused by the enemy. An American commander will call for artillery fire at the least hint of trouble. One or two enemy soldiers with automatic weapons might cause the commander who had never been told about map error to blow his whole unit away. During the errant barrage, would anyone notice a few impacts of dubious origin? When a U.S. base gets shelled, must all of its new damage necessarily be the result of the shelling?

*Secretly Occupy an Enemy-Controlled Area*

In battle, it's relatively easy to undermine and subvert a stronger adversary's power. The idea is to reverse the role of provider and recipient of an enemy-controlled variable. Literally this axiom translates, "Reverse the positions of host and guest."

> Before a battle, one can reverse the position of host and guest, and exploit a weak spot of the enemy to attack.[166]

One way is to reverse the role of occupant and occupier in an enemy-controlled area (e.g., to secretly take up residence in an enemy camp). If a forward observer cannot hide inside a defensive perimeter as it is built, he can sneak in afterwards. He needs only infiltration skills and a place to hide. Building a spider hole or subterranean room is not that difficult.

> Defeat the enemy from within by infiltrating the enemy's camp under the guise of cooperation. . . . In this way you can discover his weakness and then when the enemy's guard is relaxed, strike directly at the source of his strength.[167]

There are any number of ways the host can usurp other aspects of the guest's environment and vice versa. In Southeast Asia, the Viet Cong stole from Allied convoys much of what they needed to fight the war. The United States, on the other hand, made only halfhearted attempts to win the hearts and minds of the people.

> In Chinese military terminology, guest refers to the party that launches an attack into the territory of its enemy, and host of the party that takes up the defensive in his own territory. Though the guest holds the initiative of the attacker, he also has many problems. He has to transport provisions long distances, fight on unfamiliar ground, deal with a hostile populace, and lay siege to well-fortified cities. A skilled attacker can minimize these difficulties by raiding the enemy's base camp for supplies, befriending the local populace to recruit guides and agents, and feign weakness to lure the enemy out of his strongholds. In these ways, the guest can obtain the advantageous position of the host for himself.[168]

Then there is the psychological application of this principle —

sneaking into an enemy's psyche to cause him to believe that some ideas are his own. Of course, to succeed, the intrusion into his mind must be accomplished progressively.

> Take the opportunity to put in your foot and seize the heart of the enemy. Proceed step by step.[169]

> Whenever there is a chance, enter into the decision-making body of your ally and extend your influence skillfully step by step.[170]

It is no accident that Oriental small-unit tactical technique initially mimics its opposite. Attack looks like defense and vice versa. By allowing his foe to go first, the Asian gains the advantage.

> The strategists have a saying: I dare not play the host, but play the guest; I dare not advance an inch, but retreat a foot instead. This is known as marching forward when there is no road, rolling up one's sleeve when there is no arm, dragging one's adversary by force when there is no adversary, and taking up arms when there are no arms.[171]
> — Lao Zi

(Source: Corel Print House, *People*, Royalman)

## Stratagems for Desperate Situations

*Lull the Enemy to Sleep with Something Beautiful*

A beautiful woman can generate feelings that lower one's interest in anything else. Commonly associated emotions — vanity, lust, jealousy, envy, and hatred — can be equally disconcerting from whatever source. Literally the axiom translates, "The beauty trap."

Most at risk from this ploy is the enemy commander. His emotions will have a decided effect on his unit's chances.

> When faced with a formidable enemy, try to subdue their leader. When dealing with an able and resourceful commander, exploit his indulgence of sensual pleasures in order to weaken his fighting spirit. When the commander becomes inept, his soldiers will be demoralized, and their combat power will be greatly reduced.[172]

The "beautiful woman" in this parable can be any pleasurable emotion. Leading an opposition commander to believe that he has won, when he really hasn't, might so greatly relieve him as to qualify. Easiest to sidetrack would be those conditioned to think that they must win every battle — an expectation so unrealistic as to suggest stretching the truth. By secretly withdrawing during the final stages of a defensive stand, the Easterner could tempt an attrition-oriented attacker to overestimate the number of defenders eliminated. The Western commander who so reacted would begin to lose touch with reality. Then, with winning assured, he might place less importance on the unique circumstances of the next engagement. As the door to fabrication got wider and wider, his tactical decisions would become increasingly less viable.

> When the general becomes inept and the soldiers listless, their fighting ability will decline.[173]

Of course, the enemy's rank and file can also be manipulated. Might artificially enhancing their pride give them less reason to improve? "As dripping water wears through rock, so the weak and yielding can subdue the firm and strong."[174] By waiting for the right time to fight, the Oriental soldier unduly flatters his Western opponent.

Assist [the enemy] in his licentiousness and indulgence . . .
in order to dissipate his will.[175]
— *The Six Secret Teachings of the Tai Gong*

Increase the enemy's excesses; seize what he loves. Then
we, acting from without, can cause a response from
within.[176]
— Sima Fa, *Seven Military Classics*

*Reveal a Weakness to Cause the Enemy to Suspect a Trap*

Unusual behavior in times of crisis evokes suspicion. Placing
any doubt whatsoever in an enemy's mind creates opportunity. If
the gates to a lightly defended city (or position) were left open, a
stronger opponent might hesitate to enter, fearing a trap. His hesi-
tation could create just enough self-doubt to ensure his subsequent
defeat. Here, one must have steady nerves to succeed. Literally
the axiom translates, "The empty-city ploy."

When the enemy is superior in numbers and your situation
is such that you expect to be overrun at any moment, then
drop all pretense of military preparedness and act casu-
ally. Unless the enemy has an accurate description of your
situation, this unusual behavior will arouse suspicions.
With luck he will be dissuaded from attacking.[177]

In spite of the inferiority of your forces, deliberately make
your defensive line defenceless in order to confuse the enemy.[178]

This parable talks of leaving an undefended city's gate open
after telling its residents to keep quiet. The attacker hesitates to
enter suspecting an ambush. In the Battle for Hue City, the enemy
encouraged a few of his soldiers to remain visible in the blocks
between those heavily defended. This may have caused them to
take more fire, but probably with little effect. They were moving
from place to place so as to appear to be more numerous.

When weak, appear strong; when strong, appear weak.[179]
— Sun Zi

This axiom might also be applied to a standard operating procedure — e.g., by omitting a step to slow down the response of a suspicious foe.

Bear a weak appearance when in a weak position to create doubts in the already doubtful enemy.[180]

When your military strength is nil, deliberately appear vulnerable. The enemy will be left guessing what you're up to.[181]

*Attack the Enemy's Cohesion from Within*

How well a unit functions in war depends on how well it understands its own capabilities. By sowing doubt, confusion, and discord within the enemy's ranks, one can undermine its will to fight. Literally the axiom translates, "Turn the enemy's agents against him," or "Sow discord in the enemy's camp."

Reduce the effectiveness of your enemy by inflicting discord among them.[182]
— Sun Zi

Undermine your enemy's ability to fight by secretly causing discord between him and his friends, allies, advisors, family, commanders, soldiers, and population. When he is preoccupied settling internal disputes, his ability to attack or defend is compromised.[183]

Here the value of infiltrating the enemy camp is emphasized. When all of his strategic points, machineguns, and obstacles have been premapped, the attacker has more tactical options.

Spying is the best of all deceptive measures against the enemy. Use the enemy's spies to work for you and you will win without any loss.[184]

This axiom can also apply to starting rumors or planting false information through double agents or to allowing enemy agents to obtain false information and return to their own lines.

In adopting this stratagem, one must first know the enemy's likely response to leaked information. Make him think it's to his advantage to follow a certain course of action and exploit his mistake.[185]

During World War II, the Japanese took advantage of the way U.S. commanders reconnoiter objectives (aerial photographs) and protect strategic assets (linear defenses). On the islands about to be invaded, they erected dummy positions where Americans might have placed real ones. The U.S. preparatory fires did little damage.

Among the ways of deceiving the enemy is plunging him into a fog. Induce the enemy's spies to work for you. . . . Pretend to be unaware of the spies' activities and deliberately leak false information to them.[186]

### Feign Injury to Yourself

One who feigns injury appears to be less of a threat to his enemy. To obey the Geneva Conventions, one cannot concurrently pretend be wounded and to surrender. Literally translated, this precept says, "Inflict injury on oneself to win the enemy's trust."

People rarely inflict injuries on themselves, so when they get injured, it is usually genuine. Exploit this naivete.[187]

To be the first to gain victory, initially display some weakness to the enemy and only afterward do battle. Then your effort will be half, but the achievement doubled.[188]
— *The Six Secret Teachings of the Tai Gong*

Then there are any number of injuries that can be feigned. A command-detonated can of diesel fuel that would look to a pilot like a secondary explosion might cause a whole flight of fighter-bombers to drop their payloads in the wrong place.

The stratagem involves an element of self-sacrifice. The enemy is fooled into drawing a conclusion which is the opposite of the truth.[189]

One is reminded that there are any number of opportunities when the enemy lets down his guard.

Espionage can be conducted when the enemy takes a false injury to be genuine.[190]

## Combine Stratagems

In war, victory belongs to the commander who can launch simultaneous attacks from left and right, outside and inside, above and below. That commander can choose which attack to follow through on, while his opponent may reinforce the wrong front. Here, the literal translation is, "Interlocking stratagems."

When the enemy possesses a superior force, do not attack recklessly. Instead, weaken him by devising plots to bring him into a difficult position of his own doing.[191]

Embedded in this stratagem is the idea of continually alternating supporting and main attacks, intended feints and actual maneuvers. How victory is achieved will depend on how the defender reacts to the different initiatives.

Appearance and intention inevitably ensnare people when artfully used, even if people sense that there is an ulterior intention behind the overt appearance. When you set up ploys and opponents fall for them, then you win by letting them act on your ruse. As for those who do not fall for a ploy, when you see they won't fall for the open trap, you have another set. Then even if opponents haven't fallen for your original ploy, in effect they actually have.[192]
— Yagyu Munenori, *Family Book on the Art of War*

Do not repeat tactics which gained you victory in the past, but let your tactics be molded by the infinite variety of circumstances.[193]
— Sun Zi

Of course, a more literal translation of this axiom is also possible. For example, switching one's focus of main effort might cause his opponent to shift his formation too often and thereby tangle his forces.

> Do not engage an enemy that has many generals and numerous soldiers. Weaken its position by making its troops interlaced [entangled].194

This entanglement could occur in many ways. If two U.S. battalions were participating in a cordon operation and not completely sure of each other's location, they could be easily induced to shoot at one another. Firing at both from defiladed ground between them would almost certainly get the job done.

> If the enemy has superior strength, don't be foolhardy and engage them in battle. Instead, think of a way to entangle them for this will weaken them.195

Then more than one simultaneous deception might create additional confusion. Upon seeing through the first, the enemy will seldom suspect a second.

> A victory often results from a circumspect battle plan consisting of several interrelated ruses. . . . The first aims to reduce the enemy's maneuverability, and the second to annihilate its effective strength. . . .
> . . . Confronted with the enemy in the morning, one challenges it to battle and soon feigns defeat. After retreating for some distances, he stops and turns to challenge the enemy again, then feigns defeat and retreats. Eagerly seeking for a decision, the enemy follows in hot pursuit, with no time to take a rest. On the other hand, one has planned the repeated retreats beforehand and is therefore able to use the intervals to rest and feed the troops. At nightfall, the enemy has become tired and hungry. Feigning defeat for the last time, one scatters cooked beans on the ground. When the enemy . . . arrives, [they] stop to feed. A great victory can be achieved if one fights back at this moment.196

A ruse can also be exaggerated by combining it with tactics or obstacles. For example, the VC often would often combine the image of a peaceful village with a defensive strongpoint protected by a deep stream. A man and his wife were seen quietly cooking their food in the otherwise deserted village of "La Thap (1)," northwest of An Hoa, just before it became the final battle in the Second Tet Offensive of 1968.[197]

> In important matters one should use several strategies applied simultaneously. Keep different plans operating in an overall scheme. In this manner, if any one strategy fails, you will still have several others to fall back on. Combining even weak strategies in unison has a greater effectiveness than applying them sequentially.[198]

## Refuse to Fight

Only false pride could cause a commander to fight a battle with no strategic consequence. In the art of war, an often overlooked but highly useful talent is knowing when to run. Literally the axiom translates, "When retreat is the best option."

> If greatly outnumbered, then retreat. While it is possible for a small force to put up a great fight, in the end it will lose to superior numbers.[199]
> — Sun Zi

(Source: *FM 21-76* (1957), p. 56)

> If it becomes obvious that your current course of action will
> lead to defeat, then retreat and regroup. When your side is
> losing, there are only three choices remaining: surrender,
> compromise, or escape. Surrender is complete defeat, com-
> promise is half defeat, but escape is no defeat. As long as
> you are not defeated, you still have a chance.[200]

Under constant attack by Chiang Kai-shek's well-supplied army, Mao Tse-tung's forces retreated more than 6000 miles across China from October 1934 to October 1935. In January 1949, they defeated the Chinese Nationalist army.[201] Refusing to fight and tactically withdrawing have been successfully employed against U.S. forces at Iwo Jima, the Chosin Reservoir, and throughout the Vietnam War. It is a tactic wholly consistent with maneuver warfare doctrine.

# Notes

**SOURCE NOTES**

**Illustrations**

Pictures on pages 20, 57, 85, 125, 139, 149, 201, 221, and 233 reproduced after written assurance from Cassell PLC, London, that the copyright holder for *WORLD ARMY UNIFORMS SINCE 1939,* text by Andrew Mollo and Digby Smith, color plates by Malcolm McGregor and Michael Chappell, could no longer be contacted. The illustrations are from part I (page 95) and part II (pages 71, 90, 91, 92, and 99) of the Cassell publication. Copyright © 1975, 1980, 1981, 1983 by Blandford Press Ltd. All rights reserved.

Pictures on pages 30, 35, and 266 reproduced with permission of Stefan H. Verstappen, from *THE THIRTY-SIX STRATEGIES OF ANCIENT CHINA.* Copyright © 1999 by Stefan H. Verstappen. All rights reserved.

Picture on page 49 reproduced with permission of De Agostini UK Ltd., from *THE ARMED FORCES OF WORLD WAR II: UNIFORMS, INSIGNIA, AND ORGANISATION,* text by Andrew Mollo, color plates by Malcolm McGregor and Pierre Turner, part I, page 255. Copyright © 1981 by Orbis Publishing Ltd., London. All rights reserved.

Picture on page 50 reproduced after telephonic assurance from HarperCollins that the copyright holder for its Thomas Y. Crowell Co. imprint *GUADALCANAL,* by Irving Werstein, could no longer be contacted. The illustration is from page 20 of the HarperCollins publication. Copyright © 1963 by Irving Werstein. All rights reserved.

Picture on page 67 reproduced after telephonic assurance from Osprey Publishing Ltd., London, that this is a U.S. War Dept. (public domain) sketch in *JAPANESE ARMY OF WORLD WAR II,* Men-at-Arms Series, text by Philip Warner, color plates by Michael Youens, page 25. Copyright © 1972 by Osprey Publishing Ltd. All rights reserved.

**Text**

Reprinted with permission of HarperCollins Publishers, from
*MARSHAL ZHUKOV'S GREATEST BATTLES,* edited by Harrison E.
Salisbury. Introduction and editorial-comments copyright © 1969 by
Harrison E. Salisbury. Translation and maps copyright © 1969 by
Harper and Row Publishers, Inc. All rights reserved.

Reprinted with permission of HarperCollins Publishers, from
*HITLER'S LAST GAMBLE,* by Trevor N. Dupuy. Copyright © 1994 by
Trevor N. Dupuy. All rights reserved.

Reprinted with permission of Oxford University Press, from *THE ART
OF WAR,* by Sun Tzu, translated by Samuel B. Griffith, foreword by
B.H. Liddell Hart. Copyright © 1963 by Oxford University Press. All
rights reserved.

Reprinted with permission of Penguin Putnam Inc., from *THE GREAT
WAR AND THE SHAPING OF THE 20TH CENTURY,* by Jay Winter
and Blaine Baggett. Copyright © 1996 by Community Television of
Southern California. All rights reserved.

Reprinted with permission of Shambhala Publications, Inc., from
*THE ART OF WAR,* by Sun Tzu, translated by Thomas Cleary.
Copyright © 1988 by Thomas Cleary. All rights reserved.

Reprinted with permission of Stefan H. Verstappen, from *THE
THIRTY-SIX STRATEGIES OF ANCIENT CHINA.* Copyright © 1999
by Stefan H. Verstappen. All rights reserved.

Reprinted with permission of Naval Institute Press, from *PHASE LINE
GREEN: THE BATTLE FOR HUE,* by Nicholas Warr. Copyright
© 1997 by Nicholas Warr. All rights reserved.

Reprinted with permission of Perseus Books Group, from *MANEUVER
WARFARE HANDBOOK,* by William S. Lind. Copyright © 1985 by
Westview Press, Inc. All rights reserved.

Reprinted with permission of Lynne Rienner Publishers, Inc., Boulder,
CO, from *THE SIEGE AT HUE,* by George W. Smith. Copyright © 1999
by George W. Smith. All rights reserved.

Reprinted with permission of International Publishers Co., Inc.,
New York, NY, from MAO TSE-TUNG: AN ANTHOLOGY OF HIS
WRITINGS, edited by Anne Fremantle. Copyright © 1954, 1956 by
International Publishers Co., Inc. Copyright © 1962 by Anne
Fremantle. All rights reserved.

## ENDNOTES

### Preface

1. Richard Humble, *Marco Polo* (New York: G.P. Putnam's Sons, 1975), pp. 68, 69.
2. Marco Polo, quoted in *The Travels of Marco Polo,* revised from Marsden's translation, ed. Manuel Komroff (New York: Modern Library, 1953), pp. 44, 45.
3. *Religion and Ritual in Chinese Society,* ed. Arthur P. Wolf (Stanford, CA: Stanford Univ. Press, 1974), pp. 1-101.
4. Tou Bi Fu Tan, *A Scholar's Dilettante Remarks on War,* chapt. 10 (n.p., n.d.), quoted in *The Wiles of War,* comp. and trans. Sun Haichen (Beijing: Foreign Languages Press, 1991), p. 264.
5. "All the Kings Men," *Exxon Masterpiece Theater*, BBC in cooperation with WGBH Boston, NC Public TV, 26 November 2000.
6. Paul Begg, "Into Thin Air," in issue 31, vol. 3, *The Unexplained: Mysteries of Mind Space & Time,* from Mysteries of the Unexplained Series (Pleasantville, NY: Readers Digest Assn., 1992), p. 125.
7. Ron Moreau, "Terror Twins," *Newsweek*, 7 February 2000, p. 42.
8. David H. Reinke (Ph.D. in parapsychology and student of Buddhism), telephone conversation with author, 6 July 2000.

### Part One: The Eastern Way of War

Chapter 1: *Unheralded Success*

1. *Pakistan: A Country Study,* ed. Peter R. Blood, Federal Research Div., Library of Congress, Area Handbook Series, Hdqts. Dept. of the Army DA Pam 550-48 (Washington, D.C.: U.S. Govt. Printing Office, 1995), pp. 4-5.
2. Bruce Gilley, "Stolen Thunder," *Far Eastern Economic Review: Asian Millennium* (Hong Kong: Review Publishing, 2000), p. 94.
3. *Lets Go: India and Nepal,* ed. Nate Barksdale (New York: St. Martin's Press, 1999), p. 276.
4. Bruce I. Gudmundsson, *Stormtroop Tactics: Innovation in the German Army 1914-1918* (New York: Praeger Pubs., 1989), pp. 147-149.
5. MCI 7401, *Tactical Fundamentals,* 1st Course in Warfighting Skills Program (Washington, D.C.: Marine Corps Inst., 1989), p. 55.
6. *Senshi sosho Kantogun (1) Tai So senbi Nomonhan jiken* [Official War Hist. Series: "The Kwantung Army," vol. 1, *Preparations for the War against the USSR and the Nomanhan Incident],* ed. Boeicho boeikenshujo senshishitsu (Tokyo: Asagumo shimbunsha, 1969), p. 36,

in Leavenworth Papers No. 2, *Nomanhan: Japanese — Soviet Tactical Combat, 1939,* by Edward J. Drea (Fort Leavenworth, KS: Combat Studies Inst., U.S. Army Cmd. & Gen. Staff College, 1981), p. 19.

7. Edward J. Drea, *Nomanhan: Japanese — Soviet Tactical Combat,* 1939, Leavenworth Papers No. 2 (Ft. Leavenworth, KS: Combat Studies Inst., U.S. Army Cmd. & Gen. Staff College, 1981), pp. 18, 19.

8. Ibid., p. 88.

9. Don Moser and the editors of Time-Life Books, *China-Burma-India,* World War II Series (Alexandria, VA: Time-Life Books, 1978), pp. 101, 146-157.

10. Drea, *Nomanhan,* Leavenworth Papers No. 2, pp. 17-20.

11. Burke Davis, *Marine* (New York: Bantam Books, 1964), p. 390.

12. Gen. Vo Nguyen Giap, "Once Again We Will Win," in *The Military Art of People's War,* ed. Russel Stetler (New York: Monthly Review Press, 1970), pp. 264, 265.

13. *Mao Tse-tung: An Anthology of His Writings,* ed. Anne Fremantle (New York: Mentor, 1962), p. 69.

14. Gudmundsson, *Stormtroop Tactics,* p. 94.

15. Ibid., pp. 146, 147.

16. William S. Lind (noted historian and author of *Maneuver Warfare Handbook),* editorial comments, 1 February 2001.

17. Georgi K. Zhukov, *Marshal Zhukov's Greatest Battles,* ed. Harrison E. Salisbury, trans. Theodore Shabad (New York: Harper & Row, 1969), p. 288.

18. Gudmundsson, *Stormtroop Tactics,* p. 21.

19. Drea, *Nomanhan,* Leavenworth Papers No. 2, p. 19.

20. *Japan / Hong Kong / Singapore,* narrated by John Temple, vol. 1 of "Asian Insights Series" (Lindfield NSW, Australia: Film Australia), distributed by Films Inc. of Chicago, IL, as videocassette 0188-9019.

21. Sun Tzu, *The Art of War,* trans. Samuel B. Griffith, foreword by B.H. Liddell Hart (New York: Oxford Univ. Press, 1963), p. 82.

22. Sun Tzu, *The Art of War,* in *Sun Tzu's Art of War: The Modern Chinese Interpretation,* by Gen. Tao Hanzhang, trans. Yuan Shibing (New York: Sterling Publishing, 1990), p. 28.

23. Edwin P. Hoyt, *The Marine Raiders* (New York: Pocket Books, 1989), p. 16.

24. Ibid., p. 23.

25. "Some Interesting Facts about Korea: The Forgotten War," *DAV Magazine,* May/June 2000, p. 28.

26. DA Pamphlet 30-51, *Handbook on the Chinese Communist Army* (Washington, D.C.: Hdqts. Dept. of the Army, 7 December 1960), p. 50.

27. Ibid., p. 55.

28. Charles Soard (highly experienced infantry scout and frequent visitor to Vietnam), telephone conversation with author, 28 March 2001.

29. Terrence Maitland and Peter McInerney, *Vietnam Experience: A Contagion of War* (Boston: Boston Publishing, 1968), p. 97.

30. Trevor N. Dupuy, David L. Bongard, and Richard C. Anderson, *Hitler's Last Gamble* (New York: HarperPerennial, 1994), pp. 563, 564.

31. Ibid., pp. 185, 202, 226.

32. Andrew Mollo and Digby Smith, *World Army Uniforms Since 1939,* part 2 (Poole, England: Blandford Press, 1981), pp. 18, 19.

33. *ABC Nightly News,* 12 September 2000.

34. Memo for the record from H.J. Poole.

35. Government official, Chinese television interview, June 2000.

36. Memo for the record from H.J. Poole.

37. Gilley, "Stolen Thunder," p. 97.

38. Col. David H. Hackworth U.S. Army (Ret.) and Julie Sherman, *About Face* (New York: Simon & Schuster, 1989), p. 818.

Chapter 2: *Strategic Advantage*

1. Lucille Schulberg and the editors of Time-Life Books, *Historic India* (New York: Time-Life Books, 1968), p. 61.

2. Ibid., pp. 58, 59.

3. *The Strategic Advantage: Sun Zi & Western Approaches to War,* ed. Cao Shan (Beijing: New World Press, 1997), p. 21.

4. Sun Tzu, *The Art of War,* trans. Samuel B. Griffith, foreword by B.H. Liddell Hart, p. 82.

5. Ibid., p. vii.

6. Bajram Fana (Kosovo refugee who recently revisited his country), conversation with the author, 20 February 2001.

7. Maj.Gen. Edwin Simmons USMC (Ret.), in "An Entirely New War" segment of *Korea — the Unknown War* (London: Thames TV in cooperation with WGBH Boston, 1990), NC Public TV, n.d.

8. *The Strategic Advantage,* ed. Cao Shan, p. 102.

9. *Mao Tse-tung,* ed. Fremantle, p. 71.

10. *The Strategic Advantage,* ed. Cao Shan, pp. 46, 47.

11. Ibid., pp. 21, 22.

12. *Mao Tse-tung,* ed. Fremantle, pp. 78, 79.

13. *The Strategic Advantage,* ed. Cao Shan, p. 64.

14. Ibid., p. 2.

15. *Mao Tse-tung,* ed. Fremantle, p. 78.

16. Ibid., p. 79.

17. Ibid., p. 82.

18. Ibid., p. 100.

19. *The Strategic Advantage,* ed. Cao Shan, p. 7.

20. Ibid., p. 11.

21. Ibid., p. 49.

22. Ibid., pp. 49, 50.
23. Ibid., pp. 50, 51.
24. Michael Lee Lanning and Dan Cragg, *Inside the VC and the NVA: The Real Story of North Vietnam's Armed Forces* (New York: Ivy Books, 1992), p. 211.
25. Ibid., pp. 211, 212.
26. *The Strategic Advantage,* ed. Cao Shan, p. 21.
27. Ibid., p. 55.
28. *Mao Tse-tung,* ed. Fremantle, p. 114.
29. Truong Chinh, *Primer for Revolt,* intro. Bernard B. Fall (New York: Praeger, 1963), pp. 114-117.
30. *Sun Bin's Art of War: World's Greatest Military Treatise,* trans. Sui Yun (Singapore: Chung Printing, 1999), pp. 161, 162.
31. *The Strategic Advantage,* ed. Cao Shan, p. 56.
32. Ibid., p. 6.
33. Ibid., p. 9.


Chapter 3: *The False Face and Art of Delay*

1. Jie Zi Bing Jing, *Master Jie's Canon of War,* tome one (n.p., n.d.), quoted in *The Wiles of War,* comp. and trans. Sun Haichen (Beijing: Foreign Languages Press, 1991), p. 3.
2. *The Wiles of War,* comp. and trans. Sun Haichen (Beijing: Foreign Languages Press, 1991), pp. iii, iv.
3. Ibid., p. i.
4. *36 Stratagems: Secret Art of War,* trans. Koh Kok Kiang and Liu Yi (Singapore, Asiapac Books, 1992), p. 227.
5. Ibid., p. 225.
6. Stefan H. Verstappen, *The Thirty-Six Strategies of Ancient China* (San Francisco: China Books & Periodicals, 1999), p. 196.
7. *The Wiles of War,* comp. and trans. Sun Haichen, p. 328.
8. Liu Tao, *Six Strategies,* tome two (n.p., n.d.), quoted in *The Wiles of War,* comp. and trans. Sun Haichen (Beijing: Foreign Languages Press, 1991), pp. 285, 286.
9. Sun Tzu, *The Art of War,* trans. Samuel B. Griffith, foreword by B.H. Liddell Hart, p. vii.


Chapter 4: *The Hidden Agenda*

1. Thomas Cleary, preface to *The Art of War,* by Sun Tzu (Boston: Shambhala Pubs., 1988), pp. vii, viii.
2. *Random House Encyclopedia,* electronic ed., s.v. "Taoism."
3. *Sun Bin's Art of War,* trans. Sui Yun, pp. 161, 162.

4. *100 Strategies of War: Brilliant Tactics in Action,* trans. Yeo Ai Hoon (Singapore: Asiapac Books, 1993), p. 186.

5. *Sun Bin's Art of War,* trans. Sui Yun, p. 189.

6. Ibid.

7. Ibid., p. 190.

8. *100 Strategies of War,* trans. Yeo Ai Hoon, p. 31.

9. *Sun Bin's Art of War,* trans. Sui Yun, p. 190.

10. Ibid.

11. Ibid.

12. Ibid., p. 117.

13. Ibid.

14. Joseph C. Goulden, *Korea: The Untold Story of the War* (New York: Times Books, 1982), p. 295.

15. *100 Strategies of War,* trans. Yeo Ai Hoon, p. 17.

16. Ibid.

17. Memo for the record from H.J. Poole.

18. Memo for the record from H.J. Poole.

19. *100 Strategies of War,* trans. Yeo Ai Hoon, p. 20.

20. Ibid., p. 146.

21. Ibid., p. 195.

22. *Six Strategies for War: The Practice of Effective Leadership,* trans. Alan Chong (Singapore: Asiapac Books, 1993), pp. 91, 86.

23. Ibid., p. 119.

24. Truong Chinh, *Primer for Revolt,* pp. 186, 187.

25. Ibid., p. 185.

26. *Six Strategies for War: The Practice of Effective Leadership,* trans. Alan Chong, p. 179.

27. *Mao Tse-tung,* ed. Fremantle, pp. 132, 133.

28. Ibid., p. 137.

29. Ibid., p. 125.

30. Ibid., p. 135.

31. Ibid., p. 138.

32. Ibid., p. 133.

33. Ibid., p. 139.

34. "Some Interesting Facts about Korea," p. 28.

35. Smith, "Army Uniforms Since 1945," p. 18.

## Part Two: The Differences in Tactical Technique

Chapter 5: *Ghost Patrols and Chance Contact*

1. FMFRP 12-110, *Fighting on Guadalcanal* (Washington, D.C.: U.S.A. War Office, 1942), pp. 25, 26.

2. Hoyt, *The Marine Raiders,* pp. 111, 112.

3. Memo for the record from H.J. Poole.

4. Herbert C. Merillat, *The Island: A History of the First Marine Division on Guadalcanal August 7 — December 9, 1942* (Washington, D.C.: Zenger Publishing, 1944; reprint, Boston: Houghton Mifflin, 1979), p. 225.

5. Hoyt, *The Marine Raiders,* p. 129.

6. Ibid.

7. FMFRP 12-110, *Fighting on Guadalcanal,* p. 8.

8. Ibid., p. 35.

9. Hoyt, *The Marine Raiders,* p. 129.

10. Maj. John L. Zimmerman, *The Guadalcanal Campaign* (Washington, D.C.: Hist. Div., HQMC, 1949), p. 145.

11. Merillat, *The Island,* pp. 219, 220.

12. Attributed to Robert OBday.

13. Hoyt, *The Marine Raiders,* p. 120.

14. Merillat, *The Island,* p. 222.

15. FMFRP 12-110, *Fighting on Guadalcanal,* p. 36.

16. Hoyt, *The Marine Raiders,* p. 113.

Chapter 6: *The Obscure, Rocky-Ground Defense*

1. Bill D. Ross, *Iwo Jima: Legacy of Valor* (New York: Vintage, 1986), p. 163.

2. Ibid., pp. 72, 73.

3. Navaho codetalker, in "Japanese Codetalkers" segment of *In Search of History,* Hist. Channel, 30 March 1999.

4. Marine infantryman, quoted in *Iwo Jima: Legacy of Valor,* by Bill D. Ross (New York: Vintage, 1986), p. 135.

5. Ross, *Iwo Jima,* pp. 138, 146.

6. Merritt E. Benson, "Seventeen Days at Iwo Jima," *Marine Corps Gazette,* February 2000, p. 32.

7. Gudmundsson, *Stormtroop Tactics,* p. 49.

8. *Sands of Iwo Jima* (Hollywood, CA: Republic Pictures, 1988), videocassette.

9. Lt.Col. Whitman S. Bartley, *Iwo Jima: Amphibious Epic* (Washington, D.C.: Hist. Branch, HQMC, 1954), p. 140.

10. Ibid., p. 161.

11. Philip Warner, *Japanese Army of World War II,* Men-at-Arms Series, color plates by Michael Youens (London: Osprey Publications Ltd., 1972), pp. 23-25.

12. Ibid.

13. Ibid.

14. Ibid.

15. TM-E 30-480, *Handbook on Japanese Military Forces* (Washington, D.C.: U.S. War Dept., 1944; reprint, Baton Rouge, LA: LSU Press, 1991), pp. 143-158.

16. Ross, *Iwo Jima,* pp. 234, 250.

17. Ibid., p. 16.

18. Memo for the record from H.J. Poole.

19. Robert Sherrod, *Tarawa — The Story of a Battle* (Fredericksburg, TX: Admiral Nimitz Foundation, 1944), pp. 89, 90.

20. Michael Kernan, "'. . . [H]eavy fire . . . unable to land . . . issue in doubt,'" *Smithsonian,* November 1993, p. 128.

21. Ross, *Iwo Jima,* p. 217.

22. Col. Joseph H. Alexander, *Closing In: Marines in the Seizure of Iwo Jima,* Marines in World War II Commemorative Series (Washington, D.C.: Hist. & Museums Div., HQMC, 1994), p. 28.

23. Bartley, *Iwo Jima,* p. 48.

24. Ross, *Iwo Jima,* p. 85.

25. Bartley, *Iwo Jima,* p. 159.

26. Ross, *Iwo Jima,* p. 217.

27. Bartley, *Iwo Jima,* p. 149.

28. Alexander, *Closing In,* p. 46.

29. Ross, *Iwo Jima,* p. 217.

30. Bartley, *Iwo Jima,* p. 164.

31. Ibid., p. 149.

32. Ross, *Iwo Jima,* p. 249.

33. Bartley, *Iwo Jima,* pp. 140, 141.

34. Ibid., p. 146.

35. Ibid., p. 189.

36. Ibid., p. 186.

37. Ibid., p. 84.

38. Ross, *Iwo Jima,* pp. 243, 244.

39. Bartley, *Iwo Jima,* p. 81.

40. Ibid., p. 65.

41. Ross, *Iwo Jima,* p. 307.

42. Bartley, *Iwo Jima*, p. 131.

43. Ross, *Iwo Jima,* p. 79.

44. Bartley, *Iwo Jima,* p. 67.

45. Ibid., p. 78.

46. Ross, *Iwo Jima,* pp. 242-244.

47. Ibid., p. 307.

48. Bartley, *Iwo Jima,* pp. 172-174.

49. Ibid., p. 186.

50. TM-E 30-480, *Handbook on Japanese Forces,* p. 117.

51. Ross, *Iwo Jima,* p. 256.

52. Ibid., p. 241.

53. Bartley, *Iwo Jima,* p. 111.

54. Ross, *Iwo Jima,* pp. 293,294.

55. Bartley, *Iwo Jima,* p. 141.

56. "4th Marine Division Intelligence Report of 6 March 1945," quoted in *Iwo Jima: Amphibious Epic,* by Lt. Col. Whitman S. Bartley (Washington, D.C.: Hist. Branch, HQMC, 1954), pp. 176, 177.

57. Ross, *Iwo Jima,* p. 249.

58. Ibid., p. 305.

59. Ibid., p. 98.

60. Ibid., p. 95.

61. Ibid., p. 297.

62. Ibid., p. 262.

63. Ibid., p. 256.

64. Ibid., pp. 218, 230.

65. Ibid., p. 149.

66. 2dLt. Clifton A. Cascaden, "Selfless Devotion Immortalized by Iwo Jima Memorial," *Camp Lejeune (North Carolina) Globe,* 8 October 1999, p. 8A.

67. Ross, *Iwo Jima,* p. 283.

68. *What Cameramen Saw Through Their Lenses,* 85:00, A Tribute to WWII Combat Cameramen of Japan Series (Tokyo: Nippon TV, 1995), 3 August 1995.

69. Bartley, *Iwo Jima,* p. 15.

70. Keith Wheeler and the editors of Time-Life Books, *The Road to Tokyo,* World War II Series (Alexandria, VA: Time-Life Books, 1979), p. 42.

71. Alexander, *Closing In,* p. 5.

72. Bartley, *Iwo Jima,* p. 184.

73. Ibid., p. 149.

74. Ibid., p. 15.

75. Ross, *Iwo Jima,* p. 333.

76. Alexander, *Closing In,* p. 46.

77. Ross, *Iwo Jima,* p. 331.

78. Thomas M. Huber, "The Battle of Manila," U.S. Army Combat Studies Inst. [cited December 2002] @ cgsc.army.mil/csi/research/mout.

79. Bartley, *Iwo Jima,* pp. 127, 128.

80. H. John Poole, *One More Bridge to Cross: Lowering the Cost of War* (Emerald Isle, NC: Posterity Press, 1999), pp. 30-32.

Chapter 7: *The Human-Wave Assault Deception*

1. *The Strategic Advantage,* ed. Cao Shan, p. 55.

2. Memo for the record from H.J. Poole.

3. Eric M. Hammel, *Chosin: Heroic Ordeal of the Korean War* (Novato, CA: Presidio Press, 1981), p. 85.

4. Ibid., p. 88.

5. Ibid., pp. 99, 100.

6. Ibid., p. 89.

7. Don Campbell (Marine platoon sergeant during the bloody hill battles of 1952-53), telephone conversation with author, 26 March 2001.

8. Hammel, *Chosin,* p. 89.

9. Ibid., pp. 102-115.

10. Ibid., pp. 155, 156.

11. Ibid., p. 71.

12. Ibid.

13. Ibid., p. 71.

14. Goulden, *Korea,* p. 349.

15. Ibid., p. 358.

16. Ibid., p. 356.

17. Hammel, *Chosin,* pp. 55-59.

18. Ibid., pp. 56-59.

19. Ibid., p. 100.

20. Ibid., p. 64.

21. Lynn Montross and Capt. Nicholas A. Canzona, *The Chosin Reservoir Campaign,* U.S. Marine Operations in Korea 1950-1953, vol. III (Washington, D.C.: Hist. Branch, HQMC, 1957), p. 182.

22. FMFRP 12-110, *Fighting on Guadalcanal,* p. 22.

23. Hammel, *Chosin,* p. 59.

24. William S. Lind, *Maneuver Warfare Handbook* (Boulder, CO: Westview Press, 1985), p. 45.

25. Hammel, *Chosin,* p. 99.

26. Ibid.

27. Ibid.

28. Ibid., p. 101.

29. Montross and Canzone, *The Chosin Reservoir Campaign,* p. 159.

30. Ibid., p. 178.

31. Ibid., p. 161.

32. Hammel, *Chosin,* pp. 56-59.

33. Montross and Canzone, *The Chosin Reservoir Campaign,* p. 163.

34. Hammel, *Chosin,* pp. 56-59.

35. Ibid., pp. 64, 65.

36. Ibid., p. 99.

37. Ibid., pp. 64, 65.

38. Ibid., pp. 56-59.

39. Montross and Canzone, *The Chosin Reservoir Campaign,* p. 166.

40. Ibid., p. 161.

41. Hammel, *Chosin,* p. 87.

42. Ibid., p. 55.

43. Montross and Canzone, *The Chosin Reservoir Campaign,* p. 168.

44. Hammel, *Chosin,* p. 87.

45. Ibid.

46. Goulden, *Korea,* p. 351.

47. Memo for the record from H.J. Poole.

48. Goulden, *Korea,* p. 349.

49. Drea, *Nomanhan,* Leavenworth Papers No. 2, p. 69.

50. Hammel, *Chosin,* p. 68.

51. Ibid., p. 71.

52. Ibid., p. 68.

53. Ibid., pp. 71, 72.

54. Montross and Canzone, *The Chosin Reservoir Campaign,* p. 167.

55. Hammel, *Chosin,* p. 87.

56. Ibid., p. 85.

57. Ibid., pp. 87, 88.

58. Ibid., p. 88.

59. Ibid., p. 89.

60. Ibid., pp. 89, 90.

61. Montross and Canzone, *The Chosin Reservoir Campaign,* p. 168.

62. Ibid.

63. Goulden, *Korea,* p. 357.

64. Hammel, *Chosin,* p. 92.

65. Montross and Canzone, *The Chosin Reservoir Campaign,* p. 172.

66. Hammel, *Chosin,* p. 93.

67. Goulden, *Korea,* pp. 351, 352.

68. Montross and Canzone, *The Chosin Reservoir Campaign,* p. 174.

69. Ibid.

70. Hammel, *Chosin,* p. 90.

71. Montross and Canzone, *The Chosin Reservoir Campaign,* p. 168.

72. Hammel, *Chosin,* pp. 89, 90.

73. Montross and Canzone, *The Chosin Reservoir Campaign,* p. 174.

74. Ibid.

75. Goulden, *Korea,* p. 352.

76. Ibid., p. 356.

77. Hammel, *Chosin,* p. 98.

78. Montross and Canzone, *The Chosin Reservoir Campaign,* p. 181.

79. Hammel, *Chosin,* p. 99.

80. Ibid.

81. Ibid., pp. 99, 100.

82. Ibid., p. 101.

83. Montross and Canzone, *The Chosin Reservoir Campaign,* p. 180.

84. Hammel, *Chosin,* p. 87.

85. Goulden, *Korea,* p. 355.

86. Ibid., p. 350.

87. Hammel, *Chosin,* pp. 79, 80.

88. Goulden, *Korea,* p. 349.

89. Montross and Canzone, *The Chosin Reservoir Campaign,* p. 174.

90. Maj. Jon T. Hoffman USMCR, "The Legacy and the Lessons of Tarawa," *Marine Corps Gazette,* November 1993, p. 64.

91. DA Pamphlet 30-51, *Handbook on the Chinese Communist Army,* pp. 23, 24.

92. Attributed to Robert OBday.

93. Reg G. Grant, "Fighting the VC Way," chapter 29 of *NAM: The Vietnam Experience 1965-75* (London: Orbis Publishing Limited, 1987), p. 148.

94. Memo for the record from H.J. Poole.

Chapter 8: *The Inconspicuous, Low-Land Defense*

1. *Insight Guides: Vietnam,* ed. Helen West (Singapore: APA Publications, 1991), p. 48.

2. Ibid., pp. 45-58.

3. Bill Lind, foreword to *One More Bridge to Cross: Lowering the Cost of War,* by H. John Poole (Emerald Isle, NC: Posterity Press, 1999), p. xii.

4. Lanning and Cragg, *Inside the VC and the NVA,* pp. 223, 224.

5. Sedgwick D. Tourison, Jr., *Talking with Victor Charlie* (New York: Ivy Books, 1991), pp. 234, 235.

6. "Enemy Field Fortifications in Korea," by U.S. Army, Corps of Engineers, Intell. Div., no. 15, in *Engineer Intelligence Notes* (Washington, D.C.: Army Map Service, January 1952), pp. 2-6; and *Chinese Communist Reference Manual for Field Fortifications,* trans. Military Intell. Sect., Gen. Staff (Far East Cmd., 1 May 1951), pp. 63, 64, 112, 178; quoted in *A Historical Perspective on Light Infantry,* by Maj. Scott R. McMichael, Research Survey No. 6 (Ft. Leavenworth, KS: Combat Studies Inst., U.S. Army Cmd. & Gen. Staff College, 1987), p. 72.

7. DA Pamphlet 30-51, *Handbook on the Chinese Communist Army,* p. 19.

8. Ibid., p. 27.

9. Ibid.

10. Ibid., p. 29.

11. "Enemy Field Fortifications in Korea," by U.S. Army Corps of Engineers, no. 15, *Engineer Intelligence Notes* (Washington, D.C.: Army Map Service, January 1952), pp. 2-6; *Chinese Communist Reference Manual for Field Fortifications,* trans. Intell. Sect., Gen. Staff (Far East Cmd., 1 May 1951), pp. 63, 64, 112, 178; quoted in *A Historical Perspective on Light Infantry,* by Maj. Scott R. McMichael, Research Survey No. 6 (Ft. Leavenworth, KS: Combat Studies Inst., U.S. Army Cmd. & Gen. Staff College, 1987), pp. 85-88.

12. Maitland and McInerney, *Vietnam Experience: A Contagion of War,* p. 100.

13. Paul R. Young, *First Recon — Second to None: A Marine Reconnaissance Battalion in Vietnam, 1967-68* (New York: Ivy Books, 1992), pp. 219, 220.

14. Maj. Gary Tefler, Lt.Col. Lane Rogers, and V. Keith Fleming, Jr., *U.S. Marines in Vietnam: Fighting the North Vietnamese, 1967* (Washington, D.C.: Hist. & Museums Div., HQMC, 1984), p. 179.

15. Ibid.

16. Lanning and Cragg, *Inside the VC and the NVA,* pp. 206-208.

17. TM-E 30-480, *Handbook on Japanese Military Forces,* pp. 143-151.

18. Memo for the record from H.J. Poole.


Chapter 9: *The Absent Ambush*

1. Tefler, Rogers, and Fleming, *U.S. Marines in Vietnam: Fighting the North Vietnamese, 1967,* p. 96.

2. Keith William Nolan, *Operation Buffalo: USMC Fight for the DMZ* (New York: Dell Publishing, 1991), p. 58.

3. Ibid., p. 61.

4. Ibid., p. 68.

5. Tefler, Rogers, and Fleming, *U.S. Marines in Vietnam: Fighting the North Vietnamese, 1967,* pp. 97, 98.

6. Nolan, *Operation Buffalo,* p. 49.

7. Ibid., p. 52.

8. Ibid., p. 71.

9. Ibid., p. 68.

10. Ibid., p. 50.

11. Tefler, Rogers, and Fleming, *U.S. Marines in Vietnam: Fighting the North Vietnamese, 1967,* p. 96.

12. Nolan, *Operation Buffalo,* pp. 72, 73.

13. Tefler, Rogers, and Fleming, *U.S. Marines in Vietnam: Fighting the North Vietnamese, 1967,* p. 96.

14. Nolan, *Operation Buffalo,* p. 78.

15. Ibid., pp. 76, 77.

16. Ibid., p. 77.

17. Ibid., p. 79.

18. Ibid., p. 135.

19. Ibid., pp. 62-65.

20. Ibid., p. 65.

21. Ibid., p. 67.

22. Ibid., p. 69.

23. Ibid., p. 87.

24. Ibid., p. 93.

25. Ibid., p. 94.

26. Ibid., p. 116.

27. Ibid., p. 123.

28. Ibid., p. 153.

29. Tefler, Rogers, and Fleming, *U.S. Marines in Vietnam: Fighting the North Vietnamese, 1967,* p. 99.

30. George K. Tanham, *Communist Revolutionary Warfare* (New York: Praeger, 1967), pp. 185, 186.

Chapter 10: *The Transparent Approach March*

1. *100 Strategies of War,* trans. Yeo Ai Hoon, p. 131.

2. Lanning and Cragg, *Inside the VC and the NVA,* pp. 203, 204.

3. *NVA-VC Small Unit Tactics & Techniques Study,* ed. Thomas Pike, part I (U.S.A.R.V., n.d.), pp. I-3 — I-6.

4. Reg G. Grant, "Fighting the VC Way," pp. 148-150.

5. Lanning and Cragg, *Inside the VC and the NVA,* p. 204.

6. Memo for the record from H.J. Poole.

7. Ibid.

8. Lanning and Cragg, *Inside the VC and the NVA,* p. 203.

9. *Mao Tse-tung,* ed. Fremantle, pp. 126, 127.

10. Ibid., p. 114.

11. Ibid., pp. 128, 129.

12. Memo for the record from H.J. Poole.

13. Ibid.

14. Charles Soard (highly experienced infantry scout and frequent visitor to Vietnam), telephone conversation with author, 28 March 2001.

15. Memo for the record from H.J. Poole.

16. Ibid.

17. Lanning and Cragg, *Inside the VC and the NVA,* pp. 204, 205.

18. Memo for the record from H.J. Poole.

19. Jack Shulimson, Lt.Col. Leonard A. Blaisol, Charles R. Smith, and Capt. David A. Dawson, *U.S. Marines in Vietnam: The Defining Year, 1968* (Washington, D.C.: Hist. & Museums Div., HQMC, 1997), p. 349.

20. Ibid.

21. Memo for the record from H.J. Poole.

Chapter 11: *The Surprise Urban Assault*

1. *100 Strategies of War,* trans. Yeo Ai Hoon, p. 141.

2. Ibid., p. 162.

3. George W. Smith, *The Siege at Hue* (New York: Ballantine Publishing, 1999), p. 10.

4. Lt.Col. Robert W. Lamont, "'Urban Warrior' — A View from North Vietnam," *Marine Corps Gazette,* April 1999, p. 33.

5. Smith, *The Siege at Hue,* p. 42.

6. Ibid., p. 31.

7. Ibid., p. 36.

8. Ibid., p. 37.

9. Ibid., p. 147.

10. Ibid., p. 106.

11. Jack Shulimson et al, *U.S. Marines in Vietnam: The Defining Year,* p. 211.

12. Ibid., pp. 164-167.

13. Smith, *The Siege at Hue,* p. 31.

14. Eric Hammel, *Fire in the Streets: The Battle for Hue, Tet 1968* (Pacifica, CA: Pacifica Press, 1991), p. 35.

15. Smith, *The Siege at Hue,* p. 37.

16. Ibid., pp. 32, 33.

17. Ibid., p. 79.

18. Ibid., p. 47.

19. Ibid., p. 66.

20. Shulimson et al, *U.S. Marines in Vietnam: The Defining Year, 1968,* p. 213.

21. Smith, *The Siege at Hue,* p. 235.

22. Ibid., pp. 204-208.

23. Ibid., pp. 91, 169.

24. Shulimson et al, *U.S. Marines in Vietnam: The Defining Year, 1968,* p. 168.

25. Grant, "Fighting the VC Way," pp. 148.

26. Smith, *The Siege at Hue,* p. 121.

27. Ibid., p. 24.

28. Hammel, *Fire in the Streets,* p. 29.

29. Smith, *The Siege at Hue,* p. 82.

30. Ibid., pp. 26, 27.

31. Ibid., p. 171.

32. Hammel, *Fire in the Streets,* p. 29.

33. Shulimson et al, *U.S. Marines in Vietnam: The Defining Year, 1968,* p. 166.

34. Smith, *The Siege at Hue,* p. 89.

35. Ibid., pp. 37-40.

36. Ibid., p. 37.

37. Ibid., pp. 36-40.

38. Shulimson et al, *U.S. Marines in Vietnam: The Defining Year, 1968,* p. 167.

39. Smith, *The Siege at Hue,* pp. 46-49.

40. Ibid., pp. xiii, xiv.

41. Ibid., p. 203.

Chapter 12: *The Hidden Urban Defense*

1. Jack Shulimson, introduction to *Phase Line Green: The Battle for Hue, 1968,* by Nicholas Warr (Annapolis, MD: Naval Inst. Press, 1997), p. xviii.

2. Shulimson et al, *U.S. Marines in Vietnam: The Defining Year, 1968,* p. 203.

3. Keith William Nolan, *Battle for Hue: Tet, 1968* (Novato, CA: Presidio Press, 1983), p. 142.

4. Shulimson et al, *U.S. Marines in Vietnam: The Defining Year, 1968,* pp. 198, 199.

5. William Craig, *The Enemy at the Gates: The Battle for Stalingrad* (New York: Readers Digest Press, 1973), pp. 102, 103.

6. Hammel, *Fire in the Streets,* p. 274.

7. Ibid., p. 296.

8. Ibid., p. 350.

9. Ibid., p. 300.

10. Nicholas Warr, *Phase Line Green: The Battle for Hue, 1968* (Annapolis, MD: Naval Inst. Press, 1997), pp. 155-160.

11. Hammel, *Fire in the Streets,* p. 267.

12. A.J.P. Taylor, *The Second World War — An Illustrated History* (New York: G.P. Putnam's Sons, 1975), p. 146.

13. Shulimson et al, *U.S. Marines in Vietnam: The Defining Year, 1968,* p. 201.

14. Ibid., p. 202.

15. Nolan, *Battle for Hue,* p. 141.

16. Warr, *Phase Line Green,* pp. 104, 105.

17. Hammel, *Fire in the Streets,* p. 268.

18. Warr, *Phase Line Green,* pp. 153, 154.

19. Hammel, *Fire in the Streets,* p. 301.

20. Warr, *Phase Line Green,* pp. 113-118.

21. Ibid., p. 125.

22. Hammel, *Fire in the Streets,* pp. 304, 305.

23. Shulimson et al, *U.S. Marines in Vietnam: The Defining Year, 1968,* p. 201.

24. Ibid.

25. Warr, *Phase Line Green,* p. 126.

26. Ibid., p. 126-129.

27. Hammel, *Fire in the Streets,* p. 274.

28. Shulimson et al, *U.S. Marines in Vietnam: The Defining Year, 1968,* p. 199.

29. Hammel, *Fire in the Streets,* p. 276.

30. Warr, *Phase Line Green,* pp. 141, 142.

31. Hammel, *Fire in the Streets,* p. 281.

32. Shulimson et al, *U.S. Marines in Vietnam: The Defining Year, 1968,* p. 200.

33. Hammel, *Fire in the Streets,* p. 279.

34. Ibid., p. 299.

35. Ibid., p. 301.

36. Warr, *Phase Line Green,* pp. 141, 142.

37. Ibid., p. 156.

38. Jack Shulimson, introduction to *Phase Line Green: The Battle for Hue, 1968,* by Nicholas Warr (Annapolis, MD: Naval Inst. Press, 1997), p. xviii.

39. Warr, *Phase Line Green,* pp. 182-185.

40. Ibid., pp. 160, 161.

41. Ibid., pp. 104, 105.

42. Ibid., p. 161.

43. "5th Marine Division Intelligence Report of 24 March 1945," quoted in *Iwo Jima: Amphibious Epic,* by Lt. Col. Whitman S. Bartley (Washington, D.C.: Hist. Branch, HQMC, 1954), p. 190.

44. Warr, *Phase Line Green,* p. 171.

45. Jack Shulimson, introduction to *Phase Line Green: The Battle for Hue, 1968,* by Nicholas Warr (Annapolis, MD: Naval Inst. Press, 1997), p. xvi.

Chapter 13: *The Vanishing Besieged Unit*

1. Hammel, *Fire in the Streets,* p. 308.

2. Warr, *Phase Line Green,* p. 217.

3. Smith, *The Siege at Hue,* p. 203.

4. Ibid.

5. Memo for the record from H.J. Poole.

6. Shulimson et al, *U.S. Marines in Vietnam: The Defining Year, 1968,* p. 210.

7. Ibid., pp. 212, 213.

8. Ibid., p. 212.

9. *Insight Guides: Vietnam,* ed. Helen West, p. 204.

10. Hammel, *Fire in the Streets,* pp. 33, 34.

11. Shulimson et al, *U.S. Marines in Vietnam: The Defining Year, 1968,* p. 214.

12. Ibid., p. 213.

13. Hammel, *Fire in the Streets,* p. 340.

14. Shulimson et al, *U.S. Marines in Vietnam: The Defining Year, 1968,* p. 210.

15. "Twenty-Five Days and Nights of Continuous Fighting for the Wonderful Victory" (NVA Army, 1968), quoted in *The Siege at Hue,* by George W. Smith (New York: Ballantine Publishing, 1999), p. xv.

16. Hammel, *Fire in the Streets,* pp. 349, 350.

17. Ibid., pp. 347-348.

18. Ibid., p. 351.

19. Smith, *The Siege at Hue,* pp. 9, 169.

20. Hammel, *Fire in the Streets,* pp. 339, 340.

21. Shulimson et al, *U.S. Marines in Vietnam: The Defining Year, 1968,* p. 210.

22. Ibid., p. 352.

23. Ibid., p. 29.

24. "Twenty-Five Days and Nights of Continuous Fighting for the Wonderful Victory," quoted in *The Siege at Hue,* pp. xiii-xv.

25. Smith, *The Siege at Hue,* p. 39.

26. Shulimson et al, *U.S. Marines in Vietnam: The Defining Year, 1968,* p. 213.

27. Ibid., p. 168.

28. Douglas Pike, *PAVN: People's Army of Vietnam* (Novato, CA: Presidio Press, 1986), p. 268.

**Part Three: The Next Disappearing Act**

Chapter 14: *How Much Has War Changed?*

1. Gudmundsson, *Stormtroop Tactics,* pp. 146-175.

2. Jay Winter and Blaine Baggett, *The Great War and the Shaping of the 20th Century* (New York: Penguin Books, 1996), p. 229.

3. Gen. Shinseki, U.S. Army Chief of Staff, in "The Future of War," *Frontline,* NC Public TV, 24 October 2000.

4. Ibid.

5. Maj.Gen. Robert H. Scales, Jr., "From Korea to Kosovo," *Armed Forces Journal International,* December 1999, p. 41.

6. Ibid.

7. Lind, *Maneuver Warfare Handbook,* p. 25.

8. Lt.Gen. Arthur S. Collins, Jr., U.S. Army (Ret.), *Common Sense Training: A Working Philosophy for Leaders* (Novato, CA: Presidio Press, 1978), p. 214.

9. Max Hastings, *The Korean War* (New York: Touchstone Books, 1987), p. 196.

10. John A. English, *On Infantry* (New York: Praeger, 1981), p. 217.

11. Maj.Gen. Robert H. Scales, Jr., Commandant of the U.S. Army War College, in "The Future of War," *Frontline,* NC Public TV, 24 October 2000.

12. Dupuy, Bongard, and Anderson, *Hitler's Last Gamble,* pp. 341, 342.

13. H.J. Poole, *The Last Hundred Yards: The NCO's Contribution to Warfare* (Emerald Isle, NC: Posterity Press, 1997), pp. 225-250.

14. "Drug Wars," *Frontline,* NC Public TV, 10 October 2000.

15. Jack R. White, *The Invisible World of Infrared* (New York: Dodd, Mead & Co., 1984), p. 36.

16. Allan Maurer, *Lasers: Light Wave of the Future* (New York: Arco, 1982), p. 50.

17. Ibid., p. 17.

18. "Night Fever: NATO's 1999 Structured Technology Demonstration Showcases Night-Vision Systems," *Armed Forces Journal International,* August 1999, p. 37.

19. Ibid.

20. Ibid.

21. Ibid.

22. *Arrest and Trial,* ABC TV, 17 November 2000.

23. "Night Fever," p. 37.

24. White, *The Invisible World of Infrared,* p. 107.

25. Ibid., p. 20.

26. Maj. Jason M. Barrett, "'Turn Out the Lights, the Party's Over,'" *Marine Corps Gazette,* September 2000, p. 73.

27. Ibid., p. 76.

28. Col. William D. Siuru, Jr., USAF (Ret.), "Emerging Technologies," *Marine Corps Gazette,* January 1999, p. 45.

29. Barrett, "'Turn Out the Lights, the Party's Over,'" p. 76.

30. Ibid., p. 75.

31. Clifford L. Laurence, *The Laser Book: A New Technology of Light* (New York: Prentice Hall, 1986), pp. 178-180.

32. Scales, "From Korea to Kosovo," p. 37.

33. John Barry and Evan Thomas, "The Kosovo Cover-Up," *Newsweek,* 15 May 2000, p. 23.

34. Samuel B. Griffith, introduction to *The Art of War,* by Sun Tzu (New York: Oxford Univ. Press, 1963), p. xi.

35. Scales, "From Korea to Kosovo," p. 37.

36. Col. G.C. Thomas, 1st Marine Division Chief of Staff, quoted in FMFRP 12-110, *Fighting on Guadalcanal* (Washington, D.C.: U.S.A. War Office, 1942), p. 65.

37. Attributed to Bob OBday.

38. White, *The Invisible World of Infrared,* p. 25.

39. Hammel, *Chosin,* pp. 102-115.

40. Ibid.

Chapter 15: *The One-on-One Encounter*

1. "War Zone Mixing Mud, Chips," released by Associated Press, *Jacksonville (North Carolina) Daily News,* 18 September 2000, p. 5B.

2. Glen W. Goodman, Jr., "Revolutionary Soldier: U.S. Army's Land Warrior Program Turns the Corner and Shows Far-Reaching Potential," *Armed Forces Journal International,* October 1999, p. 56.

3. Ibid., p. 62.

4. Glen W. Goodman, Jr., "A Breed Apart: U.S. Army Rangers Train Hard to Go the Extra Mile," *Armed Forces Journal International,* October 1999, p. 88.

5. Siuru, "Emerging Technologies," pp. 40, 41.

6. Ibid., p. 41.

7. "The Future of War," *Frontline,* NC Public TV, 24 October 2000.

8. Barrett, "'Turn Out the Lights, the Party's Over,'" p. 74.

9. White, *The Invisible World of Infrared,* p. 18.

10. Ibid., p. 43.

11. Ibid., p. 41.

12. Ibid., p. 55.

13. Glenn W. Goodman, Jr., "Night-Fighting Advantage: U.S. Army Fields Next-Generation Night-Vision Devices," *Armed Forces Journal International,* December 2000, p. 45.

14. Barrett, "'Turn Out the Lights, the Party's Over,'" pp. 73, 74.

15. Buddy L. Locklear (U.S. Marine infantryman with 13 years in line units and 2 years as a tactics instructor), interview with author, 20 April 2001.

16. DA Pamphlet 30-51, *Handbook on the Chinese Communist Army,* p. 74.

Chapter 16: *America's Only Option*

1. John Hillen, Commission on National Security, in "The Future of War," *Frontline,* PBS TV, 24 October 2000.
2. "Night Fever," p. 42.
3. Kernan, "'. . . [H]eavy fire . . . unable to land . . . issue in doubt,'" pp. 121-128.
4. Goodman, "Night-Fighting Advantage," p. 46.
5. General Patton, quoted in *Hitler's Last Gamble,* by Trevor N. Dupuy, David L. Bongard, and Richard C. Anderson, Jr. (New York: HarperPerennial, 1994), pp. 498, 499.
6. Memo for the record from H.J. Poole.
7. Barrett, "'Turn Out the Lights, the Party's Over,'" pp. 71, 76.
8. White, *The Invisible World of Infrared,* pp. 117, 118.
9. Barrett, "'Turn Out the Lights, the Party's Over,'" p. 76.
10. Ibid., p. 73.


**Appendix A:** *Strategies for Deception*

1. *The Wiles of War,* comp. and trans. Sun Haichen, pp. 1, 2.
2. Ibid., p. 1.
3. *36 Stratagems,* trans. Koh Kok Kiang and Liu Yi, p. 7.
4. Musashi Miyamoto, *Book of Five Rings* (n.p., n.d.), quoted in *The Thirty-Six Strategies of Ancient China,* by Stefan H. Verstappen (San Francisco: China Books & Periodicals, 1999), p. 3.
5. Verstappen, *The Thirty-Six Strategies of Ancient China,* p. 3.
6. TM-E 30-480, *Handbook on Japanese Military Forces,* pp. 143-151.
7. TM-E 30-451, *Handbook on German Military Forces* (Washington, D.C.: U.S. War Dept., 1945; reprint, Baton Rouge, LA: LSU Press, 1990), pp. 230-238.
8. Guan Zi, *Book of Master Guan,* chapt. 17 (n.p., n.d.), quoted in *The Wiles of War,* comp. and trans. Sun Haichen (Beijing: Foreign Languages Press, 1991), p. 2.
9. Sun Zi Bing Fa, *Art of War,* chapt. 4 (n.p., n.d.). quoted in *The Wiles of War,* comp. and trans. Sun Haichen (Beijing: Foreign Languages Press, 1991), p. 3.
10. Poole, *One More Bridge to Cross,* p. 36.
11. Ibid., pp. 71, 72.
12. Sun Zi Bing Fa, *Art of War,* chapt. 6 (n.p., n.d.), quoted in *The Wiles of War,* comp. and trans. Sun Haichen (Beijing: Foreign Languages Press, 1991), p. 3.
13. Verstappen, *The Thirty-Six Strategies of Ancient China,* p. 9.
14. *The Wiles of War,* comp. and trans. Sun Haichen, p. 10.

15. *36 Stratagems,* trans. Koh Kok Kiang and Liu Yi, p. 13.

16. *The Wiles of War,* comp. and trans. Sun Haichen, p. 10.

17. Sun Zi, quoted in *The Thirty-Six Strategies of Ancient China,* by Stefan H. Verstappen (San Francisco: China Books & Periodicals, 1999), p. 9.

18. *36 Stratagems,* trans. Koh Kok Kiang and Liu Yi, pp. 25, 26.

19. Ibid., p. 20.

20. Bing Fa Bai Yan, "Borrow," from *A Hundred War Maxims* (n.p., n.d.), quoted in *The Wiles of War,* comp. and trans. Sun Haichen (Beijing: Foreign Languages Press, 1991), p. 26.

21. Verstappen, *The Thirty-Six Strategies of Ancient China*, p. 15.

22. *36 Stratagems,* trans. Koh Kok Kiang and Liu Yi, p. 27.

23. Ibid., p. 33.

24. *The Wiles of War,* comp. and trans. Sun Haichen, p. 35.

25. Sun Zi Bing Fa, *Art of War,* chapt. 6 (n.p., n.d.), quoted in *The Wiles of War,* comp. and trans. Sun Haichen (Beijing: Foreign Languages Press, 1991), p. 38.

26. *36 Stratagems,* trans. Koh Kok Kiang and Liu Yi, pp. 39, 40.

27. Verstappen, *The Thirty-Six Strategies of Ancient China*, p. 23.

28. *36 Stratagems,* trans. Koh Kok Kiang and Liu Yi, p. 34.

29. Sun Zi Bing Fa, *Art of War,* chapt. 4 (n.p., n.d.), quoted in *The Wiles of War,* comp. and trans. Sun Haichen (Beijing: Foreign Languages Press, 1991), p. 45.

30. *36 Stratagems,* trans. Koh Kok Kiang and Liu Yi, p. 47.

31. Eric Hammel, *Guadalcanal: Starvation Island* (New York: Crown Publishing, 1987), p. 221.

32. Tou Bi Fu Tan, *A Scholar's Dilettante Remarks on War*, chapt. 4 (n.p., n.d.), quoted in *The Wiles of War,* comp. and trans. Sun Haichen (Beijing: Foreign Languages Press, 1991), p. 53.

33. *36 Stratagems,* trans. Koh Kok Kiang and Liu Yi, p. 41.

34. Verstappen, *The Thirty-Six Strategies of Ancient China*, p. 34.

35. *The Wiles of War,* comp. and trans. Sun Haichen, p. 60.

36. *36 Stratagems,* trans. Koh Kok Kiang and Liu Yi, p. 50.

37. Ibid., pp. 55, 56.

38. Verstappen, *The Thirty-Six Strategies of Ancient China*, p. 31.

39. Tang Tai Zong Li Jing Wen Dui, *Li Jing's Reply to Emperor Taizong of Tang,* tome one (n.p., n.d.), quoted in *The Wiles of War,* comp. and trans. Sun Haichen (Beijing: Foreign Languages Press, 1991), p. 61.

40. *The Wiles of War,* comp. and trans. Sun Haichen, p. 68.

41. *36 Stratagems,* trans. Koh Kok Kiang and Liu Yi, p. 57.

42. Ibid., p 63.

43. Sun Zi, quoted in *The Thirty-Six Strategies of Ancient China,* by Stefan H. Verstappen (San Francisco: China Books & Periodicals, 1999), p. 35.

44. Verstappen, *The Thirty-Six Strategies of Ancient China*, p. 35.

45. Sun Zi, quoted in *The Thirty-Six Strategies of Ancient China,* by Stefan H. Verstappen (San Francisco: China Books & Periodicals, 1999), p. 41.

46. *The Wiles of War,* comp. and trans. Sun Haichen, p. 77.

47. Ibid., pp. 77, 78.

48. Bing Fa Bai Yan, "Wait," from *A Hundred War Maxims* (n.p., n.d.), quoted in *The Wiles of War,* comp. and trans. Sun Haichen (Beijing: Foreign Languages Press, 1991), p. 78.

49. *36 Stratagems,* trans. Koh Kok Kiang and Liu Yi, p. 70.

50. Verstappen, *The Thirty-Six Strategies of Ancient China*, p. 42.

51. *36 Stratagems,* trans. Koh Kok Kiang and Liu Yi, p. 71.

52. Ibid., p. 72.

53. *The Wiles of War,* comp. and trans. Sun Haichen, p. 88.

54. *The Six Secret Teachings of Tai Gong* (n.p., n.d.), quoted in *The Thirty-Six Strategies of Ancient China*, by Stefan H. Verstappen (San Francisco: China Books & Periodicals, 1999), p. 45.

55. Bai Zhan Qui Fa, "Battle of Pride," from *A Hundred Marvelous Battle Plans* (n.p., n.d.), quoted in *The Wiles of War,* comp. and trans. Sun Haichen (Beijing: Foreign Languages Press, 1991), p. 89.

56. *36 Stratagems,* trans. Koh Kok Kiang and Liu Yi, p. 78.

57. Verstappen, *The Thirty-Six Strategies of Ancient China*, p. 49.

58. Li Shimin, Emperor Taizong of the Tang Dynasty, quoted in *The Wiles of War,* comp. and trans. Sun Haichen (Beijing: Foreign Languages Press, 1991), pp. 97, 98.

59. *36 Stratagems,* trans. Koh Kok Kiang and Liu Yi, pp. 80-82.

60. Lao Zi, quoted in *The Thirty-Six Strategies of Ancient China*, by Stefan H. Verstappen (San Francisco: China Books & Periodicals, 1999), p. 49.

61. Li Su, Tang Dynasty general, quoted in *The Wiles of War,* comp. and trans. Sun Haichen (Beijing: Foreign Languages Press, 1991), p. 97.

62. *The Wiles of War,* comp. and trans. Sun Haichen, p. 105.

63. *36 Stratagems,* trans. Koh Kok Kiang and Liu Yi, p. 83.

64. *The Wiles of War,* comp. and trans. Sun Haichen, p. 106.

65. Wu Jing Zong Yao, *Summary of Military Canons*, tome three of first part (n.p., n.d.), quoted in *The Wiles of War,* comp. and trans. Sun Haichen (Beijing: Foreign Languages Press, 1991), p. 108.

66. Jiang Yan, "Grasp Opportunity," from *The Art of Generalship* (n.p., n.d), quoted in *The Wiles of War,* comp. and trans. Sun Haichen (Beijing: Foreign Languages Press, 1991), pp. 106, 107.

67. Tai Bai Yin Jing, *The Yin Canon of Vesper,* chapt. 21 (n.p., n.d.), quoted in *The Wiles of War,* comp. and trans. Sun Haichen (Beijing: Foreign Languages Press, 1991), p. 107.

68. *36 Stratagems,* trans. Koh Kok Kiang and Liu Yi, p. 84.

69. Sun Zi, quoted in *The Thirty-Six Strategies of Ancient China*, by Stefan H. Verstappen (San Francisco: China Books & Periodicals, 1999), p. 53.

70. Verstappen, *The Thirty-Six Strategies of Ancient China*, p. 59.

71. *36 Stratagems*, trans. Koh Kok Kiang and Liu Yi, p. 91.

72. Ibid., p. 90.

73. Sima Fa, *Law of Master Sima*, chapt. 5 (n.p., n.d.), quoted in *The Wiles of War*, comp. and trans. Sun Haichen (Beijing: Foreign Languages Press, 1991), pp. 116, 117.

74. Wu Bei Ji Yao, *An Abstract of Military Works* (n.p., n.d.), quoted in *The Wiles of War*, comp. and trans. Sun Haichen (Beijing: Foreign Languages Press, 1991), p. 117.

75. *Mao Tse-tung*, ed. Fremantle, p. 71.

76. Verstappen, *The Thirty-Six Strategies of Ancient China*, p. 65.

77. *36 Stratagems*, trans. Koh Kok Kiang and Liu Yi, p. 95.

78. Ibid., pp. 96, 97.

79. Ibid., p. 101.

80. Tou Bi Fu Tan, *A Scholar's Dilettante Remarks on War*, chapt. 1 (n.p., n.d.), quoted in *The Wiles of War*, comp. and trans. Sun Haichen (Beijing: Foreign Languages Press, 1991), pp. 126, 127.

81. Sun Zi, quoted in *The Thirty-Six Strategies of Ancient China*, by Stefan H. Verstappen (San Francisco: China Books & Periodicals, 1999), p. 71.

82. Verstappen, *The Thirty-Six Strategies of Ancient China*, p. 71.

83. *36 Stratagems*, trans. Koh Kok Kiang and Liu Yi, p. 102.

84. Ibid., pp. 103, 104.

85. Bai Zhan Qi Fa, "Battle of Contest," from *A Hundred Marvelous Battle Plans* (n.p., n.d.), quoted in *The Wiles of War*, comp. and trans. Sun Haichen (Beijing: Foreign Languages Press, 1991), p. 134.

86. Verstappen, *The Thirty-Six Strategies of Ancient China*, p. 76.

87. Sun Zi, quoted in *The Thirty-Six Strategies of Ancient China*, by Stefan H. Verstappen (San Francisco: China Books & Periodicals, 1999), p. 75.

88. *36 Stratagems*, trans. Koh Kok Kiang and Liu Yi, p. 107.

89. Bai Zhan Qi Fa, "Battle of Extremity," from *A Hundred Marvelous Battle Plans* (n.p., n.d.), quoted in *The Wiles of War*, comp. and trans. Sun Haichen (Beijing: Foreign Languages Press, 1991), p. 143.

90. *36 Stratagems*, trans. Koh Kok Kiang and Liu Yi, p. 111.

91. Ibid., p. 114.

92. *The Wiles of War*, comp. and trans. Sun Haichen, p. 151.

93. Ibid.

94. Bing Fa Bai Yan, "Abandon," from *A Hundred War Maxims* (n.p., n.d.), quoted in *The Wiles of War*, comp. and trans. Sun Haichen (Beijing: Foreign Languages Press, 1991), p. 153.

95. *36 Stratagems,* trans. Koh Kok Kiang and Liu Yi, p. 112.

96. Verstappen, *The Thirty-Six Strategies of Ancient China*, p. 81.

97. Sun Zi, quoted in *The Thirty-Six Strategies of Ancient China*, by Stefan H. Verstappen (San Francisco: China Books & Periodicals, 1999), p. 81.

98. Sun Zi Bing Fa, *Art of War,* chapt. 5 (n.p., n.d.), quoted in *The Wiles of War,* comp. and trans. Sun Haichen (Beijing: Foreign Languages Press, 1991), pp. 153, 154.

99. *36 Stratagems,* trans. Koh Kok Kiang and Liu Yi, p. 119.

100. Ibid., p. 123.

101. Bai Zhan Qi Fa, "Battle of Necessity," from *A Hundred Marvelous Battle Plans* (n.p., n.d.), quoted in *The Wiles of War,* comp. and trans. Sun Haichen (Beijing: Foreign Languages Press, 1991), pp. 159, 160.

102. Hu Qian Jing, *Canon of the General* (n.p., n.d.), quoted in *The Wiles of War,* comp. and trans. Sun Haichen (Beijing: Foreign Languages Press, 1991), p. 160.

103. Sun Zi Bing Fa, *Art of War,* chapt. 11 (n.p., n.d.), quoted in *The Wiles of War,* comp. and trans. Sun Haichen (Beijing: Foreign Languages Press, 1991), p. 160.

104. Verstappen, *The Thirty-Six Strategies of Ancient China*, p. 85.

105. *The Wiles of War,* comp. and trans. Sun Haichen, p. 170.

106. *36 Stratagems,* trans. Koh Kok Kiang and Liu Yi, p. 126.

107. Musashi Miyamoto, *Book of Five Rings* (n.p., n.d.), quoted in *The Thirty-Six Strategies of Ancient China*, by Stefan H. Verstappen (San Francisco: China Books & Periodicals, 1999), p. 91.

108. *36 Stratagems,* trans. Koh Kok Kiang and Liu Yi, pp. 128-130.

109. Verstappen, *The Thirty-Six Strategies of Ancient China*, p. 91.

110. Wei Liao Zi, *Book of Master Wei Liao,* chapt. 4 (n.p., n.d.), quoted in *The Wiles of War,* comp. and trans. Sun Haichen (Beijing: Foreign Languages Press, 1991), p. 172.

111. Verstappen, *The Thirty-Six Strategies of Ancient China*, p. 97.

112. *36 Stratagems,* trans. Koh Kok Kiang and Liu Yi, p. 130.

113. *The Six Secret Teachings of the Tai Gong* (n.p., n.d.), quoted in *The Thirty-Six Strategies of Ancient China*, by Stefan H. Verstappen (San Francisco: China Books & Periodicals, 1999), p. 97.

114. Tou Bi Fu Tan, *A Scholar's Dilettante Remarks on War,* chapt. 8 (n.p., n.d.), quoted in *The Wiles of War,* comp. and trans. Sun Haichen (Beijing: Foreign Languages Press, 1991), p. 179.

115. *36 Stratagems,* trans. Koh Kok Kiang and Liu Yi, p. 132.

116. Verstappen, *The Thirty-Six Strategies of Ancient China*, p. 103.

117. *36 Stratagems,* trans. Koh Kok Kiang and Liu Yi, p. 136.

118. *The Wiles of War,* comp. and trans. Sun Haichen, p. 190.

119. *36 Stratagems,* trans. Koh Kok Kiang and Liu Yi, p. 140.

120. Memo for the record from H.J. Poole.

121. *The Wiles of War,* comp. and trans. Sun Haichen, p. 189.

122. *36 Stratagems,* trans. Koh Kok Kiang and Liu Yi, p. 141.

123. Ibid., p. 142.

124. Mushashi Miyamoto, Book *of Five Rings* (n.p., n.d.), quoted in *The Thirty-Six Strategies of Ancient China,* by Stefan H. Verstappen (San Francisco: China Books & Periodicals, 1999), p. 109.

125. *The Wiles of War,* comp. and trans. Sun Haichen, p. 196.

126. *36 Stratagems,* trans. Koh Kok Kiang and Liu Yi, p. 148.

127. Verstappen, *The Thirty-Six Strategies of Ancient China,* p. 113.

128. *36 Stratagems,* trans. Koh Kok Kiang and Liu Yi, p. 152.

129. *The Wiles of War,* comp. and trans. Sun Haichen, p. 206.

130. Ibid., p. 205.

131. Verstappen, *The Thirty-Six Strategies of Ancient China,* p. 117.

132. Memo for the record from H.J. Poole.

133. *36 Stratagems,* trans. Koh Kok Kiang and Liu Yi, p. 153.

134. Ibid., p. 159.

135. Nolan, *Operation Buffalo,* p. 72.

136. Shulimson et al, *U.S. Marines in Vietnam: The Defining Year, 1968,* p. 349.

137. *36 Stratagems,* trans. Koh Kok Kiang and Liu Yi, p. 166.

138. Verstappen, *The Thirty-Six Strategies of Ancient China,* p. 123.

139. *The Wiles of War,* comp. and trans. Sun Haichen, p. 220.

140. *36 Stratagems,* trans. Koh Kok Kiang and Liu Yi, p. 163.

141. Ibid., p. 162.

142. *The Wiles of War,* comp. and trans. Sun Haichen, p. 220.

143. Verstappen, *The Thirty-Six Strategies of Ancient China,* p. 129.

144. *36 Stratagems,* trans. Koh Kok Kiang and Liu Yi, p. 171.

145. Ibid., p. 167.

146. *36 Stratagems,* trans. Koh Kok Kiang and Liu Yi, p. 172.

147. Ibid., p. 178.

148. *The Wiles of War,* comp. and trans. Sun Haichen, p. 242.

149. Bing Lei, *Essentials of War,* chapt. 9 (n.p., n.d.), quoted in *The Wiles of War,* comp. and trans. Sun Haichen (Beijing: Foreign Languages Press, 1991), p. 243.

150. Verstappen, *The Thirty-Six Strategies of Ancient China,* p. 135.

151. Ibid., p. 141.

152. *The Wiles of War,* comp. and trans. Sun Haichen, p. 252.

153. Ibid., p. 251.

154. Sun Zi, quoted in *The Thirty-Six Strategies of Ancient China*, by Stefan H. Verstappen (San Francisco: China Books & Periodicals, 1999), p. 141.

155. *36 Stratagems*, trans. Koh Kok Kiang and Liu Yi, p. 179.

156. Ibid., pp. 180, 181.

157. Nolan, *Operation Buffalo*, p. 77.

158. *36 Stratagems*, trans. Koh Kok Kiang and Liu Yi, p. 184.

159. Memo for the record from H.J. Poole.

160. *36 Stratagems*, trans. Koh Kok Kiang and Liu Yi, p. 188.

161. Verstappen, *The Thirty-Six Strategies of Ancient China*, p. 147.

162. Bing Fa Bai Yan, "Display," from *A Hundred War Maxims* (n.p., n.d.), quoted in *The Wiles of War,* comp. and trans. Sun Haichen (Beijing: Foreign Languages Press, 1991), p. 262.

163. Sun Bin, quoted in *The Thirty-Six Strategies of Ancient China*, by Stefan H. Verstappen (San Francisco: China Books & Periodicals, 1999), p. 147.

164. Tou Bi Fu Tan, *A Scholar's Dilettante Remarks on War,* chapt. 10 (n.p., n.d.), quoted in *The Wiles of War,* comp. and trans. Sun Haichen (Beijing: Foreign Languages Press, 1991), pp. 262-264.

165. Bing Lei, *Essentials of War,* chapt. 19 (n.p., n.d.), quoted in *The Wiles of War,* comp. and trans. Sun Haichen (Beijing: Foreign Languages Press, 1991), p. 262.

166. *36 Stratagems*, trans. Koh Kok Kiang and Liu Yi, p. 195.

167. Verstappen, *The Thirty-Six Strategies of Ancient China*, p. 153.

168. *The Wiles of War,* comp. and trans. Sun Haichen, p. 274.

169. Ibid., p. 273.

170. *36 Stratagems*, trans. Koh Kok Kiang and Liu Yi, p. 189.

171. Lao Zi, quoted in *The Thirty-Six Strategies of Ancient China*, by Stefan H. Verstappen (San Francisco: China Books & Periodicals, 1999), p. 153.

172. *36 Stratagems*, trans. Koh Kok Kiang and Liu Yi, p. 198.

173. Ibid., p. 199.

174. *The Wiles of War,* comp. and trans. Sun Haichen, p. 285.

175. *The Six Secret Teachings of the Tai Gong* (n.p., n.d.), quoted in *The Thirty-Six Strategies of Ancient China*, by Stefan H. Verstappen (San Francisco: China Books & Periodicals, 1999), p. 161.

176. Sima Fa, *Seven Military Classics* (n.p., n.d.), quoted in *The Thirty-Six Strategies of Ancient China*, by Stefan H. Verstappen (San Francisco: China Books & Periodicals, 1999), p. 161.

177. Verstappen, *The Thirty-Six Strategies of Ancient China*, p. 169.

178. *36 Stratagems*, trans. Koh Kok Kiang and Liu Yi, p. 203.

179. Sun Zi, quoted in *The Thirty-Six Strategies of Ancient China*, by Stefan H. Verstappen (San Francisco: China Books & Periodicals, 1999), p. 169.

180. *The Wiles of War*, comp. and trans. Sun Haichen, p. 296.

181. *36 Stratagems*, trans. Koh Kok Kiang and Liu Yi, p. 204.

182. Sun Zi, quoted in *The Thirty-Six Strategies of Ancient China*, by Stefan H. Verstappen (San Francisco: China Books & Periodicals, 1999), p. 173.

183. Verstappen, *The Thirty-Six Strategies of Ancient China*, p. 174.

184. *36 Stratagems*, trans. Koh Kok Kiang and Liu Yi, p. 208.

185. Ibid, p. 212.

186. Ibid., p. 209.

187. Ibid., p. 213.

188. *The Six Secret Teachings of the Tai Gong* (n.p., n.d.), quoted in *The Thirty-Six Strategies of Ancient China*, by Stefan H. Verstappen (San Francisco: China Books & Periodicals, 1999), p. 179.

189. *36 Stratagems*, trans. Koh Kok Kiang and Liu Yi, p. 217.

190. *The Wiles of War*, comp. and trans. Sun Haichen, p. 311.

191. *36 Stratagems*, trans. Koh Kok Kiang and Liu Yi, p. 218.

192. Yagyu Munenori, *Family Book on the Art of War* (n.p., n.d.), quoted in *The Thirty-Six Strategies of Ancient China*, by Stefan H. Verstappen (San Francisco: China Books & Periodicals, 1999), p. 185.

193. Sun Zi, quoted in *The Thirty-Six Strategies of Ancient China*, by Stefan H. Verstappen (San Francisco: China Books & Periodicals, 1999), p. 185.

194. *The Wiles of War*, comp. and trans. Sun Haichen, p. 319.

195. *36 Stratagems*, trans. Koh Kok Kiang and Liu Yi, p. 219.

196. *The Wiles of War*, comp. and trans. Sun Haichen, pp. 319, 320.

197. Memo for the record from H.J. Poole.

198. Verstappen, *The Thirty-Six Strategies of Ancient China*, p. 186.

199. Sun Zi, quoted in *The Thirty-Six Strategies of Ancient China*, by Stefan H. Verstappen (San Francisco: China Books & Periodicals, 1999), p. 193.

200. Verstappen, *The Thirty-Six Strategies of Ancient China*, p. 193.

201. DA Pamphlet 30-51, *Handbook on the Chinese Communist Army*, pp. 5-7.

# Bibliography

**Government Manuals and Chronicles**

Alexander, Col. Joseph H. *Closing In: Marines in the Seizure of Iwo Jima.* Marines in World War II Commemorative Series. Washington, D.C.: Hist. & Museums Div., HQMC, 1994.

Bartley, Lt.Col. Whitman S. *Iwo Jima: Amphibious Epic.* Washington, D.C.: Hist. Branch, HQMC, 1954.

DA Pamphlet 30-51, *Handbook on the Chinese Communist Army.* Washington, D.C.: Hdqts. Dept. of the Army, 7 December 1960.

Drea, Edward J. *Nomonhan: Japanese — Soviet Tactical Combat, 1939.* Leavenworth Papers No. 2. Ft. Leavenworth, KS: Combat Studies Inst., U.S. Army Cmd. & Gen. Staff College, 1981.

"Enemy Field Fortifications in Korea," by U.S. Army Corps of Engineers, No. 15, *Engineer Intelligence Notes.* Washington, D.C.: Army Map Service, January 1952. *Chinese Communist Reference Manual for Field Fortifications*, trans. Intell. Sect., Gen. Staff. Far East Cmd., 1 May 1951. Quoted in *A Historical Perspective on Light Infantry,* by Maj. Scott R. McMichael. Research Survey No. 6. Ft. Leavenworth, KS: Combat Studies Inst., U.S. Army Cmd. & Gen. Staff College, 1987.

FMFRP 12-110. *Fighting on Guadalcanal.* Washington, D.C.: U.S.A. War Office, 1942.

Hubler, Thomas M. *Japan's Battle for Okinawa April - June 1945.* Leavenworth Papers No. 18. Ft. Leavenworth, KS: Combat Studies Inst., U.S. Army Cmd. & Gen. Staff College, 1990.

MCI 7401. *Tactical Fundamentals.* 1st Course in Warfighting Skills Program. Washington, D.C.: Marine Corps Inst., 1989.

Montross, Lynn and Capt. Nicholas A. Canzona. *The Chosin Reservoir Campaign.* U.S. Marine Operations in Korea 1950-1953, vol. III. Washington, D.C.: Hist. Branch, HQMC, 1957.

*Pakistan: A Country Study.* Edited by Peter R. Blood, Federal Research Div., Library of Congress. Area Handbook Series. Hdqts. Dept. of the Army DA Pam 550-48. Washington, D.C.: U.S. Govt. Printing Office, 1995.

Shulimson, Jack, Lt.Col. Leonard A. Blaisol, Charles R. Smith, and Capt. David A. Dawson. *The Marines in Vietnam — 1968.* Washington, D.C.: Hist. &Museums Div., HQMC, 1998.

Tefler, Maj. Gary, Lt.Col. Lane Rogers, and V. Keith Fleming, Jr. *U.S. Marines in Vietnam: Fighting the North Vietnamese, 1967.* Washington, D.C.: Hist. & Museums Div., HQMC, 1984.

TM-E 30-480. *Handbook on Japanese Military Forces.* Washington, D.C.: U.S. War Dept., 1944. Reprint, Baton Rouge, LA: LSU Press, 1991.

TM-E 30-451. *Handbook on German Military Forces.* Washington, D.C.: U.S. War Dept., 1945. Reprint, Baton Rouge, LA: LSU Press, 1990.

Updegraph, Charles L., Jr. *U.S. Marine Corps Special Units in World War II.* Washington, D.C.: Hist. &Museums Div., HQMC, 1972.

Zimmerman, Maj. John L. *The Guadalcanal Campaign.* Washington, D.C.: Hist. Div., HQMC, 1949.


**Civilian Books, Magazine Articles, and Video/Film Presentations**

"All the Kings Men." *Exxon Masterpiece Theater.* BBC in cooperation with WGBH Boston, NC Public TV, 2 April 2000.

Ambrose, Stephen E. *Citizen Soldiers: The U.S. Army from the Normandy Beaches to the Bulge to the Surrender of Germany.* New York: Touchstone Books, 1997.

*Arrest and Trial.* ABC TV, 17 November 2000.

Barrett, Maj. Jason M. "'Turn Out the Lights, the Party's Over.'" *Marine Corps Gazette,* September 2000.

Barry, John and Evan Thomas. "The Kosovo Cover-Up." *Newsweek,* 15 May 2000.

Begg, Paul. "Into Thin Air." In issue 31, vol. 3, *The Unexplained: Mysteries of Mind Space & Time,* from *Mysteries of the Unexplained.* Pleasantville, NY: Readers Digest Assn., 1992.

Cascaden, 2dLt. Clifton A. "Selfless Devotion Immortalized by Iwo Jima Memorial." *Camp Lejeune (North Carolina) GLOBE,* 8 October 1999.

Collins, Lt.Gen. Arthur S., Jr., U.S. Army (Ret.). *Common Sense Training: A Working Philosophy for Leaders.* Novato, CA: Presidio Press, 1978.

Craig, William. *The Enemy at the Gates: The Battle for Stalingrad.* New York: Readers Digest Press, 1973.

Davis, Burke. *Marine.* New York: Bantam Books, 1964.

"Drug Wars." *Frontline.* NC Public TV, 10 October 2000.

Dupuy, Trevor N., David L. Bongard, and Richard C. Anderson Jr. *Hitler's Last Gamble.* New York: HarperPerennial, 1994.

English, John A. *On Infantry.* New York: Praeger, 1981.

Gilley, Bruce. "Stolen Thunder." *Far Eastern Economic Review: Asian Millennium.* Hong Kong: Review Publishing, 2000.

Goodman, Glen W., Jr. "Revolutionary Soldier: U.S. Army's Land Warrior Program Turns the Corner and Shows Far-Reaching Potential." *Armed Forces Journal International,* October 1999.

Goulden, Joseph C. *Korea: The Untold Story of the War.* New York: Times Books, 1982.

Grant, Reg G. "Fighting the VC Way." Chapt. 29 in *NAM: The Vietnam Experience 1965-75.* London: Orbis Publishing, 1987.

Griffith, Samuel B. Introduction to *The Art of War,* by Sun Tzu. New York: Oxford Univ. Press, 1963.

Gudmundsson, Bruce I. *Stormtroop Tactics: Innovation in the German Army 1914-1918.* Westport, CT: Praeger Pubs., 1989.

Hackworth, David H., and Julie Sherman. *About Face.* New York: Simon & Schuster, 1989.

Hammel, Eric M. *Chosin: Heroic Ordeal of the Korean War.* Novato, CA: Presidio Press, 1981.

Hammel, Eric. *Guadalcanal: Starvation Island.* New York: Crown Publishing, 1987.

Hammel, Eric. *Fire in the Streets: The Battle for Hue, Tet 1968.* Pacifica, CA: Pacifica Press, 1991.

Hastings, Max. *The Korean War.* New York: Touchstone Books, 1987.

Hoffman, Major Jon. T. USMCR. "The Legacy and the Lessons of Tarawa." *Marine Corps Gazette,* November 1993.

Hoyt, Edwin P. *The Marine Raiders.* New York: Pocket Books, 1989.

Humble, Richard. *Marco Polo.* New York: G.P. Putnam's Sons, 1975.

*Insight Guides: Vietnam.* Ed. Helen West. Singapore: APA Publications, 1991.

*Japan/Hong Kong — Singapore.* Narrated by John Temple. Vol. 1 of "Asian Insights Series." Lindfield NSW, Australia: Film Australia. Distributed by Films Inc. of Chicago, IL, as videocassette 0188-9019.

Kernan, Michael. "'. . . [H]eavy fire . . . unable to land . . . issue in doubt.'" *Smithsonian,* November 1993.

*Korea — the Unknown War.* London: Thames TV in cooperation with WGBH Boston, 1990. NC Public TV, n.d.

Lamont, Lt.Col. Robert W. "'Urban Warrior' — A View from North Vietnam." *Marine Corps Gazette,* April 1999.

Lanning, Michael Lee and Dan Cragg. *Inside the VC and the NVA: The Real Story of North Vietnam's Armed Forces.* New York: Ivy Books, 1992.

Laurence, Clifford L. *The Laser Book: A New Technology of Light.* New York: Prentice Hall, 1986.

*Lets Go: India and Nepal.* Edited by Nate Barksdale. New York: St. Martin's Press, 1999.

Lind, William S. Foreword to *One More Bridge to Cross: Lowering the Cost of War,* by H. John Poole. Emerald Isle, NC: Posterity Press, 1999.

Lind, William S. *Maneuver Warfare Handbook.* Boulder, CO: Westview Press, 1985.

Maitland, Terrence and Peter McInerney. *Vietnam Experience: A Contagion of War.* Boston: Boston Publishing, 1968.

*Mao Tse-tung: An Anthology of His Writings.* Edited by Anne Fremantle. New York: Mentor, 1962.

Maurer, Alan. *Lasers: Light Wave of the Future.* New York: Arco, 1982.

Merillat, Herbert Christian. *Guadalcanal Remembered.* New York: Donn, Mead & Co., 1982.

Merillat, Herbert C. *The Island: A History of the First Marine Division on Guadalcanal August 7 — December 9, 1942.* Washington. D.C.: Zenger Publishing, 1944. Reprint, Boston: Houghton Mifflin, 1979.

Mollo, Andrew and Digby Smith. *World Army Uniforms Since 1939.* Poole, England: Blandford Press, 1981.

Moreau, Ron. "Terror Twins." *Newsweek,* 7 February 2000.

Moser, Don and the editors of Time-Life Books. *China-Burma-India.* World War II Series. Alexandria, VA: Time-Life Books, 1978.

"Night Fever: NATO's 1999 Structured Technology Demonstration Showcases Night-Vision Systems." Armed *Forces Journal International,* August 1999.

Nolan, Keith William. *Battle for Hue: Tet, 1968.* Novato, CA: Presidio Press, 1983.

Nolan, Keith William. *Operation Buffalo: USMC Fight for the DMZ.* Dell Publishing, 1991.

*NVA-VC Small Unit Tactics & Techniques Study.* Edited by Thomas Pike, U.S.A.R.V., part I, n.d.

*100 Strategies of War: Brilliant Tactics in Action.* Translated by Yeo Ai Hoon. Singapore: Asiapac Books, 1993.

Pike, Douglas. *PAVN: Peoples' Army of Vietnam.* Novato, CA: Presidio Press, 1986.

Poole, H. John. *One More Bridge to Cross: Lowering the Cost of War.* Emerald Isle, NC: Posterity Press, 1999.

Poole, H.J. *The Last Hundred Yards: The NCO's Contribution to Warfare.* Emerald Isle, NC: Posterity Press, 1997.

*Religion and Ritual in Chinese Society.* Edited by Arthur P. Wolf. Stanford, CA: Stanford Univ. Press, 1974.

Ross, Bill D. *Iwo Jima: Legacy of Valor.* New York: Vintage, 1986.

*Sands of Iwo Jima.* Hollywood: Republic Pictures, 1988. Videocassette.

Scales, Maj.Gen. Robert H., Jr. "From Korea to Kosovo." *Armed Forces Journal International,* December 1999.

Sherrod, Robert. *Tarawa — The Story of a Battle.* Fredericksburg, TX: Admiral Nimitz Foundation, 1944.

Shulimson, Jack, Lt.Col. Leonard A. Blaisol, Charles R. Smith, and Capt. David A. Dawson. *U.S. Marines in Vietnam: The Defining Year, 1968.* Washington, D.C.: Hist. & Museums Div., HQMC, 1997.

Shulimson, Jack. Introduction to *Phase Line Green: The Battle for Hue, 1968,* by Nicholas Warr. Annapolis, MD: Naval Inst. Press, 1997.

Siuru, Col. William D., Jr., USAF (Ret.). "Emerging Technologies." *Marine Corps Gazette,* January 1999.

*Six Strategies for War: The Practice of Effective Leadership.* Translated by Alan Chong. Singapore: Asiapac Books, 1993.

Smith, George W. *The Siege at Hue.* New York: Ballantine Publishing, 1999.

"Some Interesting Facts about Korea: The Forgotten War." *DAV Magazine,* May/June 2000.

*Sun Bin's Art of War: World's Greatest Military Treatise.* Translated by Sui Yun. Singapore: Chung Printing, 1999.

Sun Tzu. *The Art of War.* Translated by Thomas Cleary. Boston: Shambhala Pubs., 1988.

Sun Tzu. *The Art of War.* Translated by Samuel B. Griffith. Foreword by B.H. Liddell Hart. New York: Oxford Univ. Press, 1963.

*Sun Tzu's Art of War: The Modern Chinese Interpretation,* by Gen. Tao Hanzhang. Translated by Yuan Shibing. New York: Sterling Publishing, 1990.

Tanham, George K. *Communist Revolutionary Warfare.* New York: Praeger, 1967.

Taylor, A.J.P. *The Second World War — An Illustrated History.* New York: G.P. Putnam's Sons, 1975.

Tefler, Maj. Gary, Lt.Col. Lane Rogers, and V. Keith Fleming, Jr. *U.S. Marines in Vietnam: Fighting the North Vietnamese, 1967.* Washington, D.C.: Hist. & Museums Div., HQMC, 1984.

"The Future of War." *Frontline.* NC Public TV, 24 October 2000.

*The Strategic Advantage: Sun Zi & Western Approaches to War.*
Edited by Cao Shan. Beijing: New World Press, 1997.

*The Travels of Marco Polo.* Revised from Marsden's translation.
Edited by Manuel Komroff. New York: Modern Library, 1953.

*The Wiles of War: 36 Military Strategies from Ancient China.*
Compiled and translated by Sun Haichen. Beijing: Foreign
Languages Press, 1991.

*36 Stratagems: Secret Art of War.* Translated by Koh Kok Kiang
and Liu Yi. Singapore, Asiapac Books, 1992.

Tourison, Sedgwick D., Jr. *Talking with Victor Charlie.* New York:
Ivy Books, 1991.

Truong Chinh. *Primer for Revolt.* Introduction and notes by
Bernard B. Fall. New York: Praeger, 1963.

"Twenty-Five Days and Nights of Continuous Fighting for the
Wonderful Victory." (NVA Army, 1968.) Quoted in *The Siege at
Hue,* by George W. Smith. New York: Ballantine Publishing,
1999.

Verstappen, Stefan H. *The Thirty-Six Strategies of Ancient China.*
San Francisco: China Books & Periodicals, 1999.

Vo Nguyen Giap, Gen. "Once Again We Will Win." In *The
Military Art of Peoples' War,* edited by Russel Stetler.
New York: Monthly Review Press, 1970.

Warner, Philip. *Japanese Army of World War II,* Men-at-Arms
Series. Plates by Michael Youens. London: Osprey
Publications Ltd., 1972.

Warr, Nicholas. *Phase Line Green.* New York: Ivy Books, 1997.

"War Zone Mixing Mud, Chips." Released by Associated Press.
*Jacksonville (North Carolina) Daily News,* 18 September 2000.

*What Cameramen Saw Through Their Lenses.* 85:00. A Tribute to
WWII Combat Cameramen of Japan Series. Tokyo: Nippon
TV, 1995.

Wheeler, Keith and the editors of Time-Life Books, *The Road to
Tokyo,* World War II Series (Alexandria, VA: Time-Life Books,
1979.

White, Jack R. *The Invisible World of Infrared.* New York: Dodd,
Mead & Co., 1984.

Winter, Jay and Blaine Baggett. *The Great War and the Shaping of
the 20th Century.* New York: Penguin Books, 1996.

Young, Paul R. *First Recon — Second to None: A Marine
Reconnaissance Battalion in Vietnam, 1967-68.* New York:
Ivy Books, 1992.

Zhukov, Georgi K. *Marshal Zhukov's Greatest Battles.* Edited by
Harrison E. Salisbury. Translated by Theodore Shabad.
New York: Harper & Row, 1969.

# About the Author

After almost 28 years as a commissioned or non-commissioned infantry officer, John Poole retired from the United States Marine Corps in April 1993. On active duty, he studied small-unit tactics for nine years: six months at the Basic School in Quantico (1966), seven months as a platoon commander in Vietnam (1966-67), three months as a rifle company commander at Camp Pendleton (1967), five months as a regimental headquarters company commander in Vietnam (1968), eight months as a rifle company commander in Vietnam (1968-69), five and a half years as an instructor with the Advanced Infantry Training Company (AITC) at Camp Lejeune (1986-92), and one year as the SNCOIC of the 3rd Marine Division Combat Squad Leaders Course (CSLC) on Okinawa (1992-93).

While at AITC, he developed, taught, and refined courses of instruction on maneuver warfare, land navigation, fire support coordination, call for fire, adjust fire, close air support, M203 grenade launcher, movement to contact, daylight attack, night attack, infiltration, defense, offensive Military Operations in Urban Terrain (MOUT), defensive MOUT, NBC defense, and leadership. While with CSLC, he further refined the same periods of instruction and developed others on patrolling.

He has completed all of the correspondence school requirements for the Marine Corps Command and Staff College, Naval War College (1000-hour curriculum), and Marine Corps Warfighting Skills Program. He is a graduate of the Camp Lejeune Instructional Management Course, the 2nd Marine Division Skill Leaders in Advanced Marksmanship (SLAM) Course, and the East-Coast School of Infantry Platoon Sergeants' Course.

Since retirement, he has continually researched the small-unit tactics of other nations and written two books. Published by Posterity Press in 1997 was *The Last Hundred Yards: The NCO's Contribution to Warfare* — a squad combat study based on the consensus opinions of 1200 NCOs and the casualty statistics of hundreds of field trials at AITC and CSLC. Then published by Posterity Press in late 1999 was *One More Bridge to Cross: Lowering the Cost of War* — a treatise on enemy small-unit tactics and how best to counter them. As of May 2001, he had conducted multiday training sessions for 30 Marine battalions (20 of them infantry) and one Naval Special Warfare Group on how to acquire common-sense warfare capabilities at the small-unit level. He has been stationed more than once in South Vietnam and Okinawa. He has visited Japan, South Korea, Taiwan, Hong Kong, Macao, Mainland China, Tibet, Nepal, India, Bangladesh, Myanmar (Burma), Thailand, Malaysia, Singapore, Indonesia, and the Philippines.

# Name Index

## A

Abrams, Creighton W.   164
An Hoa (base)   42, 145, 146, 270, 291
Antonelli, John W.   62
Assamana (village)   55

## B

Barr, John F.   160
Bataan (battle)   14
Begg, Paul   xvi
Beijing (city)   12, 28
Bell, Leon E.   131
Belleau Wood (battle)   14
Berlin (battle)   9
Bing Fa Bai Yan   245, 253, 264, 281
Bing Lei   277, 282
Bowe, Jim   168
Boyd, John   27
Budd, Talman C., II   187
Bulge (battle)   12, 82, 207, 235
Burns, Leon R.   131, 132
Butler, John A.   62

## C

Cai Be (camp)   23
Carlson, Evans   10, 51, 54, 55, 56
Chase, Jack   161
Chau Phong (village)   42, 147, 148, 270

Cheatham, Ernie   179
Chemin des Dames Offensive   202
Chi, Phan Huu   195
Chi Voi Mountain   189
Chiang Kai-shek   260, 292
Chinese Civil War of 1946-49   292
Chinese Resistance of 1937-45   44, 45, 144
Chinese Revolutions of 1926-37   17, 260, 292
Chosin Reservoir (battle)   6, 14, 89, 102, 139, 219
Citadel (battle)   150, 151, 154, 155, 156, 159, 162, 164, 166, 167, 168, 169, 171, 173, 174, 175, 176, 177, 179, 184, 186, 187, 190, 191, 193, 194, 196
Clausewitz, Karl von   17, 21, 22
Coates, Sterling K.   130, 131, 132, 133, 134
Cobb, Ty   158
Collins, Arthur S., Jr.   205
Con Thien (base)   127, 131, 137, 143
Cooke, Leroy M.   95
Cua Viet River   115
Cu Chi (district)   114

## D

Danang (base)   146, 154, 176
Delaney, William   132
Dong Ha (base)   137, 145
Downs, Michael   190